Vital Principles

For David

Vital Principles

The Molecular Mechanisms of Life

Andrew Scott

Basil Blackwell

Copyright © Andrew Scott 1988

First published 1988

Basil Blackwell Ltd
108 Cowley Road, Oxford, OX4 1JF, UK

Basil Blackwell Inc.
432 Park Avenue South, Suite 1503
New York, NY 10016, USA

British Library Cataloguing in Publication Data

Scott, Andrew *1955–*
    Vital principles
    1. Organisms. Metabolism. Biochemistry
    I. Title
    574.1 '33
ISBN 0-631-15398-5

Library of Congress Cataloging in Publication Data

Scott, Andrew, 1955–
    Vital principles.
    Bibliography: p.
    Includes index.
    1. Life (Biology)   2. Molecular biology.   I. Title.
QH501.S37   1988   574   88-6129
ISBN 0-631-15398-5

Typeset in Bembo 12 on 14pt
by Joshua Associates Ltd, Oxford
Printed in Great Britain by
T. J. Press Ltd, Padstow.

Contents

Preface

If we search within ourselves, and within all other forms of life, to discover what we are made of, we find we are made of chemicals which are composed of atoms of matter.

About 36 different types of atoms are found within the living things of earth, and they continually interact and react to create thousands of complex chemical compounds. The continual dynamic interplay of chemicals as they react to one another's presence is all we can find when we search to uncover the central mechanisms of life and to identify the principles which govern their operation. Every living thing is a stunningly complex chemical machine. Regardless of whether or not there is more to us than 'mere' chemical machinery, that machinery is certainly a vital part of us, and may well be all we are.

This book attempts to describe the essential physics, chemistry and biology of life. It will tell you what living things are made of and highlight the vital principles which underpin everything they do.

It is intended to provide a quick and easy introduction to anyone interested in how life works. It is a 'satellite view' of the living world, rather than an exhausting exploration of all its intricate features. It is a 'quick guide' rather than an atlas; a basic briefing rather than an in depth course; an attempt to 'give you your bearings' in a complex landscape, rather than show you how every contour twists and turns.

Those who are familiar with life's molecular mechanisms may gasp at some of the sweeping generalizations and simplifications needed to provide a short and readable overall view; but the book is not intended for those who know the field already. It is intended to let others gain an easy but worthwhile insight into the fascinating microworld within them.

I hope the book will be useful to a wide range of people: laypersons with an interest in science, students at school and university, scientists whose expertise lies in other fields, and perhaps even biologists and chemists interested in reminding themselves about the 'wood' whose 'trees' they know in great detail.

I have tried to assume nothing about my readers while writing this book, other than expecting them to have an interest in the subject.

Where debatable and controversial topics are covered, my aim has been to explain the 'standard view', rather than to convince anyone of its truth or to substantiate it with exhaustive evidence and examples. This is not a history book, or a textbook, or a bible. It is a summary of what scientists know, or think they know, about the molecular mechanisms of life, and, like all aspects of science, that knowledge is open to future challenge and change.

A glossary of technical terms is provided at the end of the book, along with suggestions for further reading.

The efforts of many people are required to make a book which carries only one person's name. I would like to thank everyone at Basil Blackwell for their efforts on my behalf, particularly Romesh Vaitilingam, Diyan Leake, Ruth Bowden and Pat Lawrence.

I am grateful to The Society of Authors, for an award from the Authors' Foundation which allowed me to devote more time to the project than would otherwise have been possible.

Finally, many thanks to Margaret, my wife, for her help in many ways.

Leith, Edinburgh

Introduction

Each day the earth spins in the radiant energy of the sun and something remarkable happens. Some of the energy is captured within living cells to power the creation of new life from the lifeless minerals of the world. In a feat of stunning self-regulating choreography, billions of atoms, molecules and ions become a part of the frantic chemical dance we call life. Each revolution of our planet in its stellar spotlight raises a little bit of the dust of the earth into the dance of life, while a little bit of the life crumbles back into dust.

Humanity, although mere participants in and victims of the great cycle of life and death, is able to reflect upon its meaning and attempt to uncover its inner mechanisms. For millennia we have searched for the secrets of life and dreamt of what we could do if these secrets were ours to exploit. Most members of our species have lived in times in which that search has proved fruitless, leaving only legends and the dreams. You and I, however, are living through privileged and exciting times in which the fundamental secrets of life are at last being revealed. Our knowledge of them sets us ready to create life of our own designings – indeed, in some small way that act of creation has already begun.

The simplest and most fundamental secret of life is the phenomenon of electric charge, which comes in two forms – positive and negative. These two forms of electric charge are attracted towards one another and repelled from themselves. Positive charge, in other words, is attracted to negative charge, and negative to positive; while positive charge repels other regions of positive charge, just as negative charge repels other regions of negative charge. This simple behaviour of electric charge is entitled to the status of a great secret of life, because all the chemical changes that make life live are the result of interactions between positive and negative electrical charge. The force of attraction between opposite charges and repulsion between like charges is the force that creates us. Physicists call it the electromagnetic force, but it is also the 'life force' at the heart of all life.

The electromagnetic force interacts with the energy of the world to push and pull the electrons and nuclei of atoms into a multitude of forms and combinations. In other words, it controls the chemical reactions of the atoms and ions and molecules that form both ourselves and all the other things of substance around us. Chemicals are formed and they react together because their parts are manipulated into place by the electromagnetic force; and of the myriad different types of chemicals the force creates, two are of central relevance to the chemistry of life – 'nucleic acids', and 'proteins'.

Nucleic acids, such as the well-known 'DNA', form the 'genes' that determine what we are. They are the chemical agents of heredity, which determine whether an egg cell, for example, will give rise to a marigold, a mosquito, a mouse or a man; and these chemical agents of heredity are copied and passed down through the generations in ways that ensure that subsequent generations of organisms are similar, though often subtly different from, the creatures that gave rise to them.

How do mere chemicals called nucleic acids organize the world around them into all manner of living things? In essence, they do it automatically, pushed and pulled by the electromagnetic force in ways that result in the creation of other chemicals – the proteins – whose structure depends on the structure of the nucleic acids that gave rise to them. Once proteins are formed, the electromagnetic force ensures that they promote the thousands of complex chemical reactions that turn lifeless raw materials into living animals and plants.

So there are three great secrets of life, which can be stated so succinctly they would occupy only a few lines in a research scientist's notebook:

The electromagnetic force interacts with the energy of the world to make chemical reactions proceed.

Chemicals called nucleic acids are able to form and to direct their own reproduction and also to direct the production of chemicals called proteins.

Proteins, directly or indirectly, promote the specific chemical reactions needed to create and sustain all life.

If, while leafing through a scientist's notebook, you chanced upon such a boldly terse summary of the central principles of life, I suspect you would become curious to know precisely what it meant and how it could possibly be justified. This book attempts to satisfy that curiosity.

1 The mystery

We all share a very humble origin – we all started life as a single cell about one tenth of a millimetre in diameter. That is very big for a cell, but obviously tiny compared with the vast and complicated creatures we quickly become. When a human egg cell begins to divide and create a newborn child it achieves an enlargement equivalent to a lightbulb giving rise to a massive office block 250 metres high; which then, over the next 15 years or so, stretches and widens to an astounding 1,000 metres in height and nearly 250 metres across. In the 'office block' that is you all the plumbing, heating, lighting, telecommunication and ventilation systems were assembled automatically and work together smoothly to sustain a bewildering diversity of very different 'suites' and 'offices'.

Human beings are really far more complex and organized objects than office blocks, so a consideration of the problems involved in getting giant new office blocks to assemble automatically gives only the merest hint of the problems involved in creating humans from fertilized egg cells. Yet this apparent miracle is completed thousands of times each day throughout the world, and similar miracles create all manner of simpler creatures, from elephants and birds and flies to bacteria and flowers and mighty oaks.

The creation of a thinking human from a tiny fertilized egg cell presents biologists with their greatest challenge. They must explain, first, the complex chemistry that allows the egg to live and grow; and, secondly, the way in which that first cell can quickly multiply into an organized adult which can move and eat and see and talk and think.

That challenge has not yet been completely overcome. The origin of consciousness and thought remains a great dilemma about which we have only a few very vague and probably simplistic ideas. Many details of the programme of development that turns an egg cell into an adult remain unknown. But we think we know most of the essential details of how cells and even such complex multicellular creatures as ourselves actually manage to live.

Inside the cell

Fertilized egg cells are rather specialized cells, having been created from the fusion of two other unusual types of cell – the sperm and egg cells. They do, however, share many general characteristics with all animal cells, and it is these general characteristics which we will explore first. As we do so, you may meet some unfamiliar chemical and biological terms. This should not worry you at this stage, since they will all be explained later in the book. Read this chapter as a general overview of the mysteries and problems whose solutions will occupy us for the rest of the book. If you cannot wait for the explanations and definitions, however, remember that there is a glossary of the key technical terms at the back of the book.

Figure 1.1 is a very simplistic view of an animal cell, showing all its essential components. Remember, though, that all cells are three-dimensional globular structures, so the view of figure 1.1 represents a thin cross-section through a cell. Real cells can be spherical, star-shaped, long and tubular, thin and flat, or of indetermined irregular and variable shapes. Although most living cells are invisible to the naked eye, they are real tangible three-dimensional objects which can be seen quite easily through microscopes. You should remember, however, that, like ourselves, they are composed of 'mere chemicals'. Each of the structures illustrated in figure 1.1 is constructed out of much smaller atoms and molecules, as we will see later.

Cells are often thought of as 'bags of water', with lots of chemicals dissolved in the water and with lots of larger structures 'floating about' in it. This is a very simplistic picture, but it does convey one essential fact about the chemistry of the cell – it occurs in a watery (aqueous)

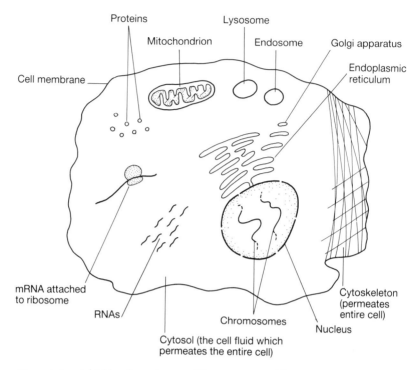

Proteins   Lysosome

Mitochondrion   Endosome   Golgi apparatus

Cell membrane   Endoplasmic reticulum

mRNA attached to ribosome   Cytoskeleton (permeates entire cell)

RNAs

Chromosomes   Nucleus

Cytosol (the cell fluid which permeates the entire cell)

Figure 1.1   A highly schematic view of the most essential features of an animal cell.

environment. Water, in other words, is the solvent of life, meaning that it is the liquid which permeates into all the nooks and crannies of the cell and in which all the chemical reactions of life take place. There are various small regions of the cell from which water is excluded, especially within the interior of some large molecules; but the chemistry of life largely proceeds in an ocean of water. It is not a clear ocean – thousands of different types of chemical are dissolved in it, and it is criss-crossed by a dense tangle of giant molecules which form 'fibres' or 'cables' or 'scaffolding' throughout the cell. Swimming through the cell 'cytosol' (the internal 'fluid' of the cell) would be like struggling through a dense underwater forest of seaweed, or through a thick paste or jelly, rather than darting through clear ocean. Nevertheless, water is virtually everywhere within the cell, and its chemical properties govern almost everything which happens there.

The boundary of all cells is formed by a thin membrane composed mainly of fatty chemicals known as 'lipids'. It is rather similar in structure to the membrane that forms a bubble of soap or detergent, although stabilized and strengthened in various ways to make it much less likely to burst. This outer membrane encloses a number of other internal membrane-bound objects. These are like small bubbles within the bigger bubble of the cell membrane itself, and some of their names might be familiar to you. The best known of these 'bubbles' within the cell is the 'nucleus' – a large dense body found within all cells. It is inside the nucleus that we find our DNA, in the form of long string-like structures studded with proteins and known as 'chromosomes'.

The other membrane-bound bodies shown in figure 1.1 are the 'mitochondrion' (of which there are usually many per cell), the 'endoplasmic reticulum', the 'Golgi apparatus' and 'endosomes' and 'lysosomes'. I will eventually be telling you something about the activities of each of them, but for the moment we can forget about them and concentrate on the essentials. The essentials really involve only four of the structures shown in the figure, composed of two broad classes of chemical. The four structures are the chromosomes, the RNAs, the ribosomes and the proteins; and all four are composed either of chemicals known as nucleic acids, or of proteins, or of a combination of the two.

Most people know that the DNA within the nucleus of their cells exists as some sort of 'double-helix' and contains the 'instructions' or the 'information' that makes them what they are, so we will start from there. As I have already said, the DNA exists in the form of long string-like double-helical shaped molecules surrounded by molecules of the other essential type of chemical, the proteins. It is this combination of DNA and protein in the nucleus that is known as a chromosome.

The DNA within any chromosome contains many hundreds of distinct regions which are known as 'genes'. So a gene is simply a specific region or 'stretch' of a longer DNA molecule. One gene occupies one region of a chromosome containing many genes, much like one song occupies one region of a music tape containing many songs overall (see figure 1.2). The importance of most genes is that they give rise to particular protein molecules, in a way that is summarized in figure 1.2.

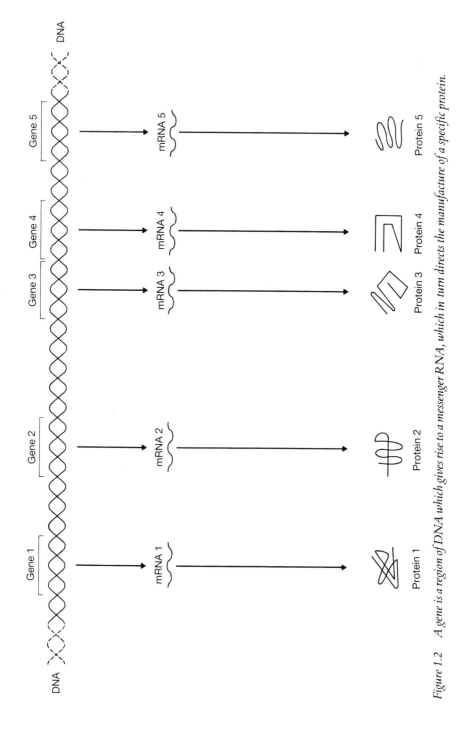

Figure 1.2   A gene is a region of DNA which gives rise to a messenger RNA, which in turn directs the manufacture of a specific protein.

First of all, in ways which we will eventually explore, a copy of each gene is made. This copy is not an exact copy since, for one thing, it is not made of DNA. Instead it is made of a very similar chemical known as 'RNA'. Both DNA and RNA belong to the same class of chemicals known as the nucleic acids, whose structure will be revealed to you in chapter 3. Secondly, while the DNA of genes is in the form of double-stranded or double-helical molecules, their RNA copies are single strands (they are actually copies of just one strand of the double-helical DNA that gave rise to them).

The RNA copies of genes are known as messenger RNAs ('mRNAs') and they travel out from the nucleus and into the cytosol. Once in the cytosol they bind to complexes of protein and RNA called 'ribosomes', and it is then that each mRNA gives rise to a protein molecule. A ribosome travels down its attached mRNA, a bit like a bead running down a thread (or sometimes like a thread being pulled through a bead), and as it does so it causes a series of simple chemicals called 'amino acids' to be linked up into a larger molecule of protein. Each protein is a unique chemical, depending on the type of amino acids it contains and the sequence in which they are arranged. So each gene in the nucleus of a cell eventually causes a specific protein molecule to appear in the cell cytosol. The importance of genes is that they give rise to proteins (plus a few important RNAs). This is the same thing as saying that the importance of DNA is that it gives rise, via RNA intermediates, to proteins; and, since both DNA and RNA are nucleic acids, we can also say that the importance of nucleic acids is that they give rise to proteins.

I have just described the central essential mechanism by which life manages to live. It lives because molecules of DNA manage to direct the production of other molecules of protein. So what do proteins do that makes them so powerful and vital? Proteins could be described as the molecular workers which actually construct and maintain all cells, and they do this job of cell construction and maintenance in essentially rather simple ways.

The most vital task which proteins perform is to act as chemical 'catalysts' which speed up specific chemical reactions. The reactions they accelerate are the very ones required to convert chemical raw materials into all the various parts of living cells, and without the assistance of proteins these reactions would proceed at rates that would be so slow as to be virtually negligible. Proteins which perform these vital acts of catalysis are called 'enzymes'. Virtually all the complex chemical reactions proceeding within living cells are catalysed by enzymes; and, since cells and all living things are the end result of all these chemical reactions, enzymes are the molecules that assemble lifeless atoms into life.

Enzymes are not the only proteins that matter though. Other proteins serve important structural roles, as the intracellular 'scaffolding' called the 'cytoskeleton', for example (see figure 1.1), or as the muscle fibres that allow us to move. Some proteins sit in the cell membrane and allow various raw materials to enter the cell, or wastes to pass out. Other proteins act as freight vehicles transporting various chemicals around the body, or as chemical messages which are sent from one cell to another. All these diverse functions of the proteins are explored in chapter 4, but, in the meantime, we are ready to summarize the basics of how cells are made and manage to live.

All cells contain molecules of DNA containing many genes. Each gene manages to bring about the manufacture of a specific protein molecule via a copy of itself which consists of the related nucleic acid known as RNA. The protein molecules produced or 'encoded' by genes go on to construct and maintain the structure of all cells, by acting as enzymes which catalyse all the required chemical reactions, and in various other ways. So cells can live because they contain genes that encode the proteins needed to make them live. The form and activities of any cell are determined by the proteins it contains, and this in turn is determined by the genes it contains. The differences between all organisms, from mice and men to magnolias and meningococci bacteria, are determined by which genes, and therefore which proteins, they contain.

Before returning to the specific example of a fertilized human egg cell, to see how it gives rise to an adult human, I should point out one other vital aspect of the activities of proteins and DNA. Some of the proteins which DNA gives rise to are able to duplicate or 'replicate' the DNA of a cell to yield two complete copies of all the DNA, and therefore of all the genes, where previously there was only one. This replication process allows cells to multiply by gradually growing and then dividing in two, because it produces a copy of all the genes needed by the cell (its 'genome') for each of the two new cells it gives rise to. This ability of DNA to become replicated, assisted by the catalytic powers of enzymes, explains why our genes are the molecular agents of heredity, ensuring that life can continue from generation to generation of very similar creatures. Just as each new pair of cells formed by cell division contains genes derived from the original 'parent' cell, so the genes that make you what you are have been derived from your own parents (although in a rather more complicated way). Genes composed of DNA are the chemicals that are 'handed down' from one generation of life to the next, ensuring that life continues on its endless course.

That completes our first cursory appraisal of the essential components and activities of living cells. If it is all new to you, it may have left you feeling somewhat confused and bewildered, wondering what all these 'things' (DNA, RNA, proteins, membranes, etc. . . .) really are, and how they can possibly turn mere lifeless chemicals into such a creature as yourself. These are the problems that are tackled by this book as a whole, so you should not expect to arrive at the answers just yet. If it was all new to you, you should perhaps read it over again to consolidate the 'general drift' in your mind before moving on. If it was an easy reminder of things you used to know, or perhaps a boring regurgitation of things you are well aware of, then it is time to move back to the fertilized egg, and on.

An egg becomes an embryo, becomes an adult

You consist of at least many billions, and perhaps even trillions, of individual cells, all derived from a single fertilized egg cell by a series of cell divisions. One cell can grow and eventually split into two in a process known as 'mitosis', during which the DNA of the original cell is replicated and then a copy passed on to each of the two new 'daughter' cells. These two cells can grow and then both divide to yield a total of four cells, and so on. You might think that an enormously long series of such synchronous divisions would be required to give rise to all the cells that make you, but calculation indicates otherwise (see figure 1.3). A few moments with a calculator will quickly convince you

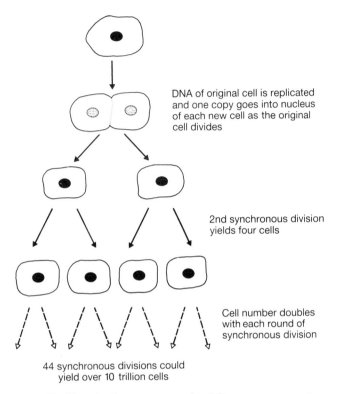

DNA of original cell is replicated and one copy goes into nucleus of each new cell as the original cell divides

2nd synchronous division yields four cells

Cell number doubles with each round of synchronous division

44 synchronous divisions could yield over 10 trillion cells

Figure 1.3   Cells able to divide in two possess the ability to generate massive numbers of cells very easily.

that only 44 generations of synchronous cell divisions would yield more than ten trillion cells from one original cell, and ten trillion is one of the higher estimates of the number of cells in a human. (Just punch in '1', press '×2 =' 44 times, and see from the display how fast your 'embryo' grows!)

To begin with, the human embryo does enlarge through a series of synchronous cell divisions, into two cells, four cells, eight cells, and so on. Eventually, however, the synchrony is lost as some regions of the embryo begin to divide faster than others and some cells stop dividing altogether. Nevertheless, the calculation indicates that, given an original cell capable of growing and dividing, we face no problem in explaining how it can yield the number of cells required to form such massive multicellular creatures as ourselves.

If cell division were all that happened during development, then what would be produced? Obviously it would simply yield a mass of identical cells essentially no different from the fertilized egg cell they had come from. Here lies the great challenge of embryology, because a newborn human is certainly not an amorphous mass of identical cells. It is an exquisitely sculpted organism composed of several hundred different types of cell, arranged to form a diversity of organs and tissues such as a heart and circulatory system, a liver, a stomach, two kidneys, two eyes; and the nervous system, lungs, muscles and bones that combine to produce the kicks and screams with which it announces its arrival.

Essentially, the problem facing scientists trying to explain the origin of any adult organism from its fertilized egg cell is as follows. The egg cell contains the entire genome of the organism, in other words it contains all the genes needed to direct the production of all the protein molecules that all the different types of cells in the adult will ever need. Obviously, as the egg cell develops into an embryo and then an adult, the cells must all become specialized to perform their eventual roles, as blood cells, liver cells, brain cells or whatever; and this specialization must be brought about by the particular proteins needed to form each particular type of cell. So how does this happen?

One way might be for cells gradually to lose all the genes that they do not need, so that brain cells would be left with only the genes needed to make brain cells, and so on. That is not what happens. It is believed that every living cell of your body contains an intact copy of the entire human genome (apart from red blood cells, which lose their nuclei in the final stages of their development). The answer to the problems of cell specialization seems to depend on the ability of genes to be 'switched on' and 'switched off' as required. A gene which is switched on, or 'activated' is one which is actively being copied into messenger RNAs which give rise to the protein encoded by the gene. A switched off or 'inactivated' gene is in a resting state in which none of the protein it codes for is being made. So in a brain cell the genes needed by an adult brain cell are active, while all the other genes of the human genome are inactive.

The development of any one type of cell is not a simple question of switching on, say, the brain cell genes. Instead, cells pass through a series of different developmental states as they change from 'un-differentiated' (i.e. unspecialized) early embryo cells into fully 'differentiated' (specialized) cells such as brain, blood or muscle cells. At each stage in the differentiation of a cell a different battery of genes must be active, to make the proteins needed by the cell at that stage and needed to move the cell on to the next stage. Some so-called 'housekeeping' genes probably must be active in all types of cells, but others may be needed only once in the development of only one type of cell.

Obviously then the development of a human from its egg cell is a very complex and co-ordinated process involving many intricate networks of gene control, which themselves must be controlled either by genes that code for gene-controlling proteins, or by various changes taking place in the environment; and yet this incredibly complicated and accurate process of development appears to proceed automatically, as a result of 'mere chemicals' reacting with one another as they are pushed and pulled by the electromagnetic force.

The development of undifferentiated cells into the right types of differentiated cells is not the only process the chemical reactions of the cell bring about. They also ensure that the various types of cells are made in the right numbers at the right places and at the right times, so that a human with two arms and two legs and a head and body all the correct size is produced, rather than a random jumble of arms and legs and eyes and hair and teeth.

Somehow the awesome developmental process which can turn a cell into a human usually proceeds without problems. The human then lives out its life for 70 years or so, thanks to the continuing chemical activities of the myriad cells from which it is made. Eventually, however, things begin to go wrong. The cells and then the whole organism begin to degenerate until some ultimate catastrophe brings to a quick end the life that the cells have sustained for so long. Many of the cells can continue to live for a while after we would pronounce the entire organism 'dead', but eventually they also die. The once proud 'pinnacle of evolution', as we like to refer to ourselves, becomes the chemical raw material needed to sustain newly emerging life. A human life has arisen from a cell, lived out its days, and then gone.

The same fate befalls all the individual creatures of the earth, although the phenomenon of life rolls endlessly on as each generation spawns the next before it goes.

The living cell; its growth and reproduction; and the organization and specialization of many living cells to create the complex multi-cellular creatures of the modern world – these are the central mysteries which our three secrets of life – life's three most vital principles – must explain.

2 Simple facts

If you take the trouble to examine complex things and events in detail, you usually find that their causes are essentially much simpler than you first supposed. A few simple principles and phenomena can produce an amazing complexity when they operate together in many different combinations and situations.

All science consists of an attempt to find the basic simplicities which underlie the complexity of the world around us. Countless scientists through the ages have known the feeling of satisfaction and surprise that comes with the discovery that something which seemed impossibly complex is really the effect of a combination of some basically simple events driven by a few simple principles.

As you delve deeper into the workings of Nature, many things actually become simpler, rather than harder, to understand. It may be much harder at first to work out what the underlying simplicities are, compared with cataloguing the complexities; but once they are discovered the simplicities make the understanding of Nature much easier.

Chemistry involves far fewer 'basic phenomena' (electrons, protons, neutrons, atoms, energy, etc.) than those that appear as the 'basic phenomena' of biological science (species, organisms, organs, tissues, cells, membranes, nuclei, mitochondria, genes, enzymes, etc., etc.). Fundamental physics has the smallest number of basic phenomena – perhaps only space-time, matter, charge, force and energy, perhaps even less.

At the biological level, you are a very complex thing indeed; but, right down at the level of basic physics and chemistry, that complexity may arise from just the five fundamental things or phenomena of physics mentioned above: space-time, matter, charge, force and energy. That bold statement assumes that you are 'only' a purely physical phenomenon, constructed solely from the chemicals that can be found within you. Many people believe that to be so, but many others are not so sure, especially those who believe in some sort of 'God' and some sort of 'soul'. For most of this book I will proceed on the assumption that all living things are simply the end result of the physics and chemistry within them. Even if the true situation is not that simple, the physical and chemical events within us are at the very least a crucial part of what we are.

Matter – the stuff of life

Of the five basic phenomena listed above, matter is the easiest to understand, at least at a superficial level. Matter is the 'stuff' from which you are made, and from which all the other things around you are made. You can slap your hand against matter, in the form of a brick wall, for example, and feel sure that it is there. You can lift it up and roll it about in the palm of your hand. You can change its shape and fashion it into all sorts of useful tools and machines with which to change and control other bits of matter. Matter appears to be nice sensible 'common-sense' solid stuff, unlike so many of the abstract ideas of physics; but nobody really knows what matter is, and when you begin to explore its structure in more detail you eventually find its reassuring solidity vanishing into ghostly abstraction.

The term 'stuff' is no less meaningful than the term 'matter', despite the latter's ring of scientific certainty. Matter is simply the stuff that we find around us and within us – the 'bits' from which everything is made, and as scientists began to probe deeper into matter they discovered that all bits of matter are composed of collections of a few smaller, simpler bits, which are themselves composed of other smaller, simpler bits. . . .

If we could shrink down to explore the microworld of chemistry and physics within us, we would eventually begin to see giant assemblies composed of many roughly spherical bits of matter all bonded together in some way (see figure 2.1). These assemblies are known as 'molecules' and the individual 'spheres' they are constructed from are 'atoms'. Molecules can be massive structures of many thousands of atoms all bonded together, such as DNA or protein molecules, or they can be composed of just two or three bonded atoms.

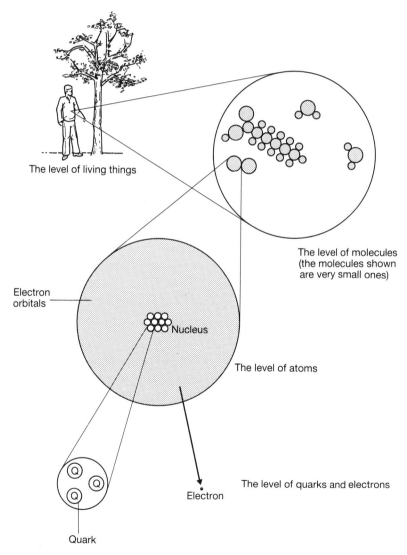

The level of living things

The level of molecules
(the molecules shown
are very small ones)

Electron
orbitals

Nucleus

The level of atoms

The level of quarks and electrons

Electron

Quark

Figure 2.1   The different levels of chemical structure within living things.

There are many millions of different types of naturally occurring molecules on earth, but they are all composed of collections of atoms selected from a repertoire of 92 different types. These atoms are listed in the 'periodic table' (figure 2.2), which becomes so drearily familiar to us throughout our schooldays. Despite the dreariness induced by gazing at a usually faded wallchart through seemingly endless days at school, the periodic table is really a very exciting and fascinating document. It lists the 92 naturally occurring atoms (plus a few man-made ones) which are the basic raw materials of chemistry – the little bits of stuff which combine in endless variety to make us work.

Atoms, of course, are themselves composed of still smaller bits of matter. If we could shrink down past the level of individual atoms we would soon discover that each atom contains a central core or nucleus, surrounded by incredibly tiny bits of matter – the sub-atomic particles called 'electrons'. The electrons carry a negative electrical charge, while the nucleus carries a positive charge. If we examined a few nuclei closely we would find them to contain two types of particles – 'protons', carrying a positive electrical charge, and 'neutrons', carrying no overall charge. We would then see that the electrical charge on each proton, designated as $+1$, is exactly equal but opposite to the charge on each electron, designated as -1. Each atom would be found to contain equal numbers of protons and electrons, making each atom electrically neutral (i.e. having no electrical charge) overall.

If our shrinkage continued, we would discover that the protons and neutrons contained yet another level of inner structure – the level of the 'quarks'. Protons, for example, are composed of three different types of the tiny particles scientists have called quarks. These quarks are rather bizarre particles with fractional electric charges ($+$ or $-\frac{2}{3}$ or $\frac{1}{3}$) and nobody has ever detected a quark directly, their existence has merely been inferred from a series of complex experiments and observations.

Modern science knows of no further levels of structure within quarks or within electrons. Quarks and electrons may be truly 'fundamental' bits of matter, with no internal structure at all, or they may not.

To understand how living things manage to live, we do not have to

Periodic table of the elements

Legend:
- Atomic number (no. of protons)
- Symbol
- Mass (in atomic mass units)

1	2	3	4	5	6	7	8	9	10	11	12	13	14	15	16	17	18
1 H 1																	2 He 4
3 Li 7	4 Be 9											5 B 11	6 C 12	7 N 14	8 O 16	9 F 19	10 Ne 20
11 Na 23	12 Mg 24											13 Al 27	14 Si 28	15 P 31	16 S 32	17 Cl 35.5	18 Ar 40
19 K 39	20 Ca 40	21 Sc 45	22 Ti 48	23 V 51	24 Cr 52	25 Mn 55	26 Fe 56	27 Co 59	28 Ni 59	29 Cu 64	30 Zn 65	31 Ga 70	32 Ge 73	33 As 75	34 Se 79	35 Br 80	36 Kr 84
37 Rb 85	38 Sr 88	39 Y 89	40 Zr 91	41 Nb 93	42 Mo 96	43 Tc 98	44 Ru 101	45 Rh 103	46 Pd 106	47 Ag 108	48 Cd 112	49 In 115	50 Sn 119	51 Sb 122	52 Te 128	53 I 127	54 Xe 131
55 Cs 113	56 Ba 137	57 La 139	72 Hf 178.5	73 Ta 181	74 W 184	75 Re 186	76 Os 190	77 Ir 192	78 Pt 195	79 Au 197	80 Hg 201	81 Tl 204	82 Pb 207	83 Bi 209	84 Po 210	85 At 210	86 Rn 222
87 Fr 223	88 Ra 226	89 Ac 227															

Lanthanides and actinides:

58 Ce 140	59 Pr 141	60 Nd 144	61 Pm 147	62 Sm 150	63 Eu 152	64 Gd 157	65 Tb 159	66 Dy 162	67 Ho 165	68 Er 167	69 Tm 169	70 Yb 173	71 Lu 175
90 Th 232	91 Pa 231	92 U 238	93 Np 237	94 Pu 242	95 Am 243	96 Cm 247	97 Bk 247	98 Cf 251	99 Es 254	100 Fm 253	101 Md 256	102 No 254	103 Lw 257

Elements beyond atomic no. 92 do not occur naturally on earth

Figure 2.2 The period table of the elements.

worry about quarks. We can get along fine with only the protons, neutrons and electrons that should be at least vaguely familiar to most people from their schooldays. To understand the chemistry of life we do not even have to worry about protons and neutrons very much. These are found clustered together at the heart of all atoms, one proton for each electron in the atom, and a variable number of the neutral neutrons. The things which interact and move about to make chemistry work are the electrons; but to understand how interacting and moving electrons can give rise to life we need to look more closely at how the electrons are arranged within the atoms of which they are part.

Atoms and their electrons

Mankind's ideas about atoms have undergone a series of radical changes through the years, and may change some more in the future (see figure 2.3). Atoms were once regarded as tiny indivisible spheres, until their inner structure of electrons, protons and neutrons was discovered. At one time it was thought that such 'sub-atomic particles' were simply mixed together like the fruit of a plum pudding, until the protons and neutrons were discovered to be clustered together into a tiny central nucleus, while the electrons were supposed to 'whirl' about the nucleus like satellites orbiting the earth, or planets circling the sun.

This planetary model of atomic structure made the small-scale world within us seem pleasingly consistent with the larger universe all around, but unfortunately it is wrong. Modern science has revealed that electrons do not seem to follow neatly defined orbits around the nucleus. The idea of orbital order has been replaced with a new concept of electrons which can never be firmly pinned down to any one particular location or orbital path. Certainly, electrons appear to surround the atomic nucleus, but in a way that allows them to dart to and fro in a seemingly chaotic manner within a particular region of space. The region of space in which a particular electron might be found is still called the electron's 'orbital', but an electron orbital is a three-dimensional volume of space, rather than a single circular orbit; and although some orbitals are spherical, others are far more bizarre.

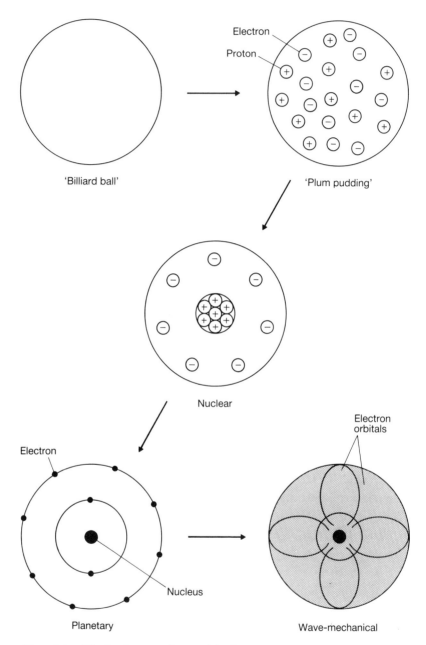

Figure 2.3   The development of our models of atomic structure.

Despite these modern complications, the essential facts about atomic structure remain quite simple. Atoms are composed of protons, neutrons and electrons. The protons and neutrons are found clustered together in a central nucleus, which is positively charged due to the protons. The negatively charged electrons are found throughout a much larger volume of space surrounding the nucleus, and they behave as if they are constantly on the move. Atoms are electrically neutral overall, since the number of positively charged protons in the nucleus is equal to the number of negatively charged electrons surrounding the nucleus, and the charge on a proton is exactly equal to, though opposite in sign from, the charge on an electron.

A few moments thought about that summary should uncover a dilemma. Electrons are negatively charged while the protons of the nucleus are positively charged. One of the most widely known principles of physics is that objects carrying opposite electrical charges are attracted towards one another, so why do the negatively charged electrons not fall in towards the positively charged nucleus due to the pull of the electromagnetic force which attracts opposite charges towards one another? Exploring the answer to the dilemma will lead us away from this discussion of matter alone, and on to consider three of the other fundamental phenomena responsible for our lives – charge, force and energy.

Force and antiforce

Everybody learns at school about the electromagnetic force, which pulls positive charge towards negative charge, and causes like charges to be repelled, but nobody really knows why either the charges or the force should exist. Objects which 'feel' or 'respond to' the electro-magnetic force are said to carry electric 'charge', but that is simply a form of words used to describe what is ultimately a mysterious phenomenon which we can describe, but not really explain.

The electromagnetic force is one of only four 'fundamental forces' at work in the universe. Gravity is one of the others, it is the force which attracts all matter in the universe towards all other matter. We can easily observe this to be the case, and we attempt to explain it by saying that all matter feels a 'gravitational force' attracting it to all other matter, but we do not know why that force should exist.

The remaining two, and least well-known, of the fundamental forces are called the 'strong nuclear force' and the 'weak nuclear force'. The strong nuclear force is the force which binds protons together in the nuclei of atoms, resisting the force of electromagnetic repulsion which would otherwise cause the protons to fly apart. This strong force is obviously crucial to chemistry, and therefore biology, since without it atoms as we know them could not exist, but it has little direct part to play in the changes that make all chemical reactions work, which is why it is unfamiliar.

The weak nuclear force governs some subtle interactions and changes within atomic nuclei which are involved in the radioactive disintegration of the nuclei, but which really need not concern us.

It used to be thought that there were other fundamental forces at work in the universe. The force of magnetism, for example, which brings North and South magnetic poles together and pushes like magnetic poles apart, used to be thought of as a fundamental force. But in the mid-nineteenth century the Scottish physicist James Clerk Maxwell proved that the phenomena of magnetism and electricity were inextricably linked, and that one unified force – the electromagnetic force – was responsible for both the attraction and repulsion of electric charges and of magnetic poles.

In recent years it has become apparent that the weak nuclear force may also be merely another manifestation of electromagnetism, so it may be more correct to talk of only three fundamental forces, rather than four.

Some physicists believe that the process of unifying the fundamental forces will eventually be taken further. They feel that gravity, electromagnetism and the strong and weak nuclear forces might well be different manifestations of one universal 'superforce' whose unity was apparent only in the earliest days after the 'big bang'. That remains to be seen, but for the moment even three or four forces represent a stunning simplification. Everything that happens appears to be made to happen by, at most, four forces straining at the particles of matter that make up the universe.

You might think of many of the other forces that we see and feel around us – the force of the tide or the force of the wind, the biological force within our arms and legs that allows us to stand upright and walk about and move things from place to place, the force of water rushing down a mountainside, or the force of an explosion blasting out the rock of a quarry. All such forces, if they are examined closely, can be shown to be effects of one or some combination of the four fundamental forces – gravity, electromagnetism, and the strong and weak nuclear forces. These few fundamental forces are the forces that direct all activity in the universe, pushing and pulling and changing things to make the universe into what it is.

How does this knowledge help to resolve our dilemma – the dilemma of why the negatively charged electrons surrounding their positively charged atomic nuclei do not go crashing into these nuclei, pulled by the electromagnetic force, to destroy all atoms, all chemistry, and all life? A discussion of the fundamental forces does not directly help us to resolve the dilemma, but it does clear the ground for a look at the phenomenon which will resolve it – the phenomenon of 'energy'.

Everybody has heard about energy, but few have thought deeply about what it really is. Energy is really an idea invented by mankind, rather than some definite thing. You can get a good intuitive grasp of what the concept of energy means by considering some situations which are described as either 'high energy' or 'low energy' states. A rock on top of a mountain is in a high energy state – it is being pulled downwards by the gravitational force, yet it is not as far down the mountainside as it could be. Two positively charged metal spheres being held against one another are in a high energy state – they would naturally be pushed apart by the force of electromagnetic repulsion, yet they are kept close together by our hands, or some other restraint. A positively charged sphere nestling into a negatively charged sphere is a much lower energy state – the electromagnetic force is pulling the spheres together, and they are as close together as they could possibly be. If they were kept apart, they would be in a higher energy state.

The clue is that high energy states always seem to be associated with some resistance against or violation of a fundamental force. Low energy states are associated with compliance with the fundamental forces. This implies that energy can be thought of as some sort of 'force resistance' or 'antiforce' able to counteract the pushes or pulls of the fundamental forces.

Armed with this definition we can say that the electrons surrounding an atomic nucleus must possess some energy, sufficient to resist the effects of the force of electromagnetic attraction pulling them towards the nucleus. Electrons stay in their orbitals around atomic nuclei because they have the energy needed to stay there, just as satellites stay in orbit around the earth because a rocket has provided them with the energy needed to do so.

Energy comes in three fundamental and interconvertable forms – potential energy, kinetic energy and radiant energy. Potential energy is the sort of energy possessed by a body whose *position* involves resisting the effects of some force, so an electron some distance from the positively charged atomic nucleus has potential energy, as does a spring which is held coiled up against the force that tries to relax it. Kinetic energy is the energy associated with *moving* objects (a form of energy which can be squared with the idea of energy as antiforce because, on colliding with something, a moving object can 'push against' or cause some 'violation of' a fundamental force). Radiant energy consists of rays of energy travelling through space, such as the rays of light and other forms of 'electromagnetic radiation' (gamma rays, x-rays, infrared rays, radio-waves, etc.).

Different forms of energy can be interconverted in many situations. Some of the energy held by an electron occupying some high energy orbital around an atomic nucleus can be given out as radiant electromagnetic energy if the electron falls down to a lower energy orbital. Conversely, electrons can absorb electromagnetic radiation, in the form of sunlight, for example, to be 'kicked up' into higher energy orbitals. Some of the energy stored in the electronic structure of 'high energy' chemicals can be released to the environment as the kinetic energy of heat during the course of a chemical reaction. In other words, as chemicals react to reach lower energy states the energy they lose can serve to speed up the motions and vibrations of the chemicals concerned and of other chemicals around them ('heat' is just a measure of the kinetic energy with which particles of matter are moving). When heat energy is used to make a chemical reaction proceed it can serve to push the chemicals involved up into unstable states of higher potential energy.

You may find many other forms of energy listed in textbooks of science, such as 'sound energy', 'nuclear energy', 'chemical energy' and 'electrical energy', but when these forms of energy are examined closely they are all found to be due to the potential energy of direct force resistance, the kinetic energy of motion, or pure radiant energy in one of its guises.

No matter how hard you try to simplify it and illustrate its effects, energy remains an abstract phenomenon which many people find hard to understand. I should say that it remains an abstract *idea* because, as I have already indicated, energy is really an idea invented by the human mind to try to make sense of the world it lives in. Just as it has made sense to suppose that matter is composed of lots of tiny particles, so it makes sense to suppose that the idea of energy as defined above corresponds closely to something which really exists in the microworld within and all around us. Force and charge are two other ideas which have proved of great benefit in helping us to make sense of, and use of, the world around us.

We are now venturing towards the top of the slippery slope that leads down to endless philosophical debate. The purpose of this book is not to explore the dark depths of theoretical physics, it is to see how the current ideas of physics and chemistry allow us to make sense of the phenomenon of life. So, rather than delving further into the real meaning of our concepts of matter, charge, force and energy, let us accept them for what they seem, and move on into the world of chemical reactions.

Chemistry - atoms interacting

What has been said so far has presented a picture of a universe which is composed of atoms of matter which are subject to the pushes and pulls of the various fundamental forces, while the universe contains a certain amount of energy able to resist these fundamental forces. Electrons occupy various orbitals around the nuclei of atoms, each electron having the appropriate energy to keep it in the orbital it is in.

The atoms of the universe do not sit aloof and forever apart, however. For one thing, they are all constantly on the move since they all possess a certain amount of the 'energy of movement' known as kinetic energy. This causes them to be constantly bumping into and bouncing off one another. When they do bump into one another something extremely significant can sometimes happen – they can interact, or 'react', to become linked up into molecules.

So what happens when atoms react together to form molecules, and why does it happen? Consider the case of the simplest atoms of all – hydrogen atoms. A hydrogen atom consists of a single proton surrounded by a solitary electron (see figure 2.4). The electron occupies an orbital in which its energy is just sufficient to balance the electromagnetic force pulling it towards the nucleus. If you were given a sample of hydrogen gas, however, you would find that instead of consisting of lots of individual hydrogen atoms, it consists entirely of hydrogen *molecules*, each molecule being composed of two hydrogen atoms bonded together. In this case we say the atoms in the molecule are 'bonded' together because their two electrons appear to be *shared* between them. Each electron is in an orbital surrounding both nuclei, rather than only one.

This is how molecules are formed. They are formed when their

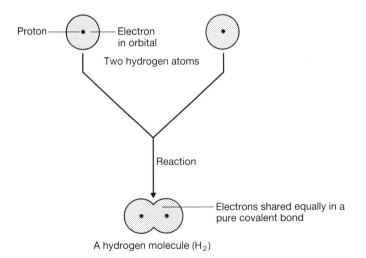

Figure 2.4   Molecules form when electrons become shared between atoms.

constituent atoms come together to leave at least some of their electrons shared between them. In the case of a hydrogen molecule, the shared orbital (the 'molecular orbital') is distributed evenly between the two atomic nuclei. This is because the two nuclei are identical, so the electrons experience an equal attraction to them both. In many molecules, however, this is not the case.

In a molecule of water, for example, the electrons involved in bonding the atoms together are much more strongly attracted to the nucleus of the oxygen atom, which has eight protons, than they are to the solitary protons that form the nuclei of the hydrogen atoms. This means that the shared electrons 'spend more time' around the oxygen atom than they do around the hydrogen atoms (see figure 2.5). This makes the oxygen atom slightly negatively charged (denoted δ^-) relative to the hydrogen atoms, since the oxygen atom gains an 'extra share' of electrons over and above the share needed to neutralize the positive charge of its nucleus. Conversely, the hydrogen atoms will carry a slight positive charge (denoted δ^+), since they are being deprived of some of the negative charge needed to neutralize the positive charge of their nuclei.

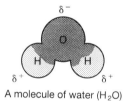

A molecule of water (H_2O)

Figure 2.5   In many molecules electrons are shared unequally between atoms in polar covalent bonds.

Hydrogen molecules and water molecules (and all other molecules) are held together by virtue of the fact that electrons are *shared* between the individual atoms involved, a similarity recognized by saying that in such cases the atoms are held together by 'covalent' bonds. In the case of water, however, the bonds are further described as 'polar' covalent bonds, since the unequal sharing of electrons between different atoms causes a polarization of electric charge.

There is a third way in which individual atoms can react together to form chemical compounds (a 'compound' is any chemical composed of two or more atoms chemically bonded together). I can illustrate this alternative using the example of sodium chloride (common salt). When sodium atoms react with chlorine atoms electrons are actually *transferred* from one atom to the other (see figure 2.6). One electron which is relatively loosely held by a sodium atom can move over to become attached to a chlorine atom. This, of course, means that the sodium atom will be left with an overall positive charge, having lost one of the electrons needed to neutralize the positive charge of its nucleus; while the chlorine atom will have acquired an overall negative charge, since it will have gained one electron over and above the number needed to neutralize the positive charge of its nucleus. In such a state the sodium 'atom' is in fact no longer a true atom at all, it has become a sodium 'ion'. The chlorine atom, on the other hand, has become a chloride ion. By definition, an ion is formed when an atom or a molecule loses or gains one or more electrons to leave it with an overall electrical charge.

The sodium and chloride ions formed in this reaction obviously have opposite electrical charges. This causes them to be attracted towards one another until they become held together by the force of electromagnetic attraction in what is called an 'ionic bond'.

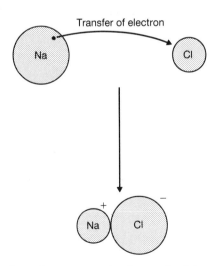

Figure 2.6   Some chemical reactions involve the transfer of electrons, creating positive and negative ions which are drawn together into an ionic bond.

The last few paragraphs have summarized some very basic chemical principles which you may have forgotten long ago, or may never have learned. Essentially, they tell us that when atoms come into contact with one another they can react to become bonded together in various ways. If they become bonded together by virtue of electrons being shared between the atoms involved, then the bonds which link them are called covalent. If identical atoms are bonded in this way the electrons will be shared equally between the atoms in a 'pure' covalent bond. If different types of atom are bonded together by electron sharing, the electrons will be more strongly attracted to the atoms whose nuclei have the largest positive charge. This situation of unequal electron distribution gives rise to polar covalent bonds, and leaves some parts of the molecules with a slight positive charge and others with a slight negative charge.

In some cases, however, atoms can react by transferring electrons between themselves to yield positively charged and negatively charged ions. These ions are then attracted towards one another, by virtue of their opposite electrical charges, to become electromagnetically 'stuck' together in an ionic bond.

When molecules or ions are brought into contact with one another, they too can react to end up in different configurations, with different bonds holding them together in new ways. That, essentially, is what happens in all chemical reactions: atoms and/or molecules and/or ions come into contact with one another, causing chemical bonds to be broken and/or made to form new chemicals composed of different combinations of atoms, molecules and ions. In all cases what is really happening when chemicals react is that some of their electrons are being redistributed between the chemicals involved.

Having grasped that, the next obvious question is why? Why do two hydrogen atoms react together to form a hydrogen molecule? Why do hydrogen molecules and oxygen molecules react to yield molecules of water? And so on. Although you might be perfectly happy to accept that atoms and molecules and ions *can* react together to cause redistributions of their electrons, nothing I have said so far can explain why such reactions should actually take place. The answer, like the answer to our earlier dilemma concerning the structure of atoms, lies with the phenomenon of energy.

To understand why chemicals should react with one another you first have to accept and understand something absolutely crucial about energy. I can summarize it in six words: energy disperses towards an even distribution; but to explain it will take a few more.

Think, first of all, of a familiar effect of energy which everyone accepts without trouble. Suppose you are holding the end of a metal fork and you place the prongs near the flames of a fire (perhaps you are making toast the old fashioned way). The far end of the fork gets hot, because heat energy is passing into it from the fire; and then eventually the end you are holding will also get hot, causing you to remove the fork from the flames, or drop it, as the heat energy passes into your skin to heat it up and cause pain.

What has happened? Very simply, energy has *dispersed* from a region where there is a lot of it (the flames of the fire) to regions where there is less (the tip of the fork, then its handle, then your skin). We all know that heat energy moves inevitably from hot places to cold places, and that it will never spontaneously move in the opposite direction, but when we experience that and accept it as plain common sense we are unlikely to realize immediately that it reveals to us the fundamental driving force of the universe.

The everyday example of heat energy not only makes it easy to appreciate that energy does disperse towards an even distribution, it also makes it easy for us to see why. Remember that when we say something is 'hot', or has a lot of 'heat energy', all we are really saying is that the particles it is made up of are moving about faster than the particles of cooler things. When a fast-moving particle bumps into a slower-moving one it inevitably causes the slower-moving particle to bounce off moving faster than before, while the fast-moving one bounces off moving a bit slower than before. Energy, in other words, is inevitably transferred from fast-moving particles to slow-moving ones when they collide. To be convinced of that, think of what happens when a fast-moving pool ball hits a stationary (or slow-moving) one. Both balls will bounce off one another, but the fast-moving one will have slowed while the other will have speeded up.

So heat energy inevitably disperses towards an even distribution because it has no other choice. As fast-moving atoms at the hot end of a piece of metal jostle about and bump into slower-moving ones, they inevitably speed up the slower-moving ones, slow down themselves, and so cause heat to travel through the metal. Eventually, the heat will become evenly distributed throughout the piece of metal; and then the metal will slowly cool as its heat energy is transferred to the cooler air around it (due to collisions between the atoms of the metal and the molecules of the air).

What goes for heat energy also holds for energy in any other guise. Energy will always move from regions of high energy into regions of low energy, whenever there is a means of energy transfer available for it to do so, and it does so because it 'has no choice'. That is essentially all that is said by the great second law of thermodynamics, the most important scientific law of them all. The example above should have convinced you of the truth of that law in one circumstance, similar examples would convince you of its truth in all other circumstances.

The second law of thermodynamics is often stated using the unfamiliar and difficult concept of 'entropy'. In this form the law states that 'all changes involve an increase in the entropy of the universe', with entropy being loosely defined as the 'degree of disorder' of the universe. This is essentially another way of saying that all changes involve the dispersal of energy. Whenever a state which we would describe as an ordered one changes into one which we would describe as more disordered, the change is driven by the automatic dispersal of energy from an uneven distribution towards a more even one. When heat moves through a piece of metal, it is moving from a so-called 'ordered' state, in which most of the heat is collected together at one end, into a 'disordered' state in which the heat is distributed throughout the metal. When a gas seeps out of a leaking cylinder into the air of a room, it is moving from an ordered state, in which it is all gathered together within the cylinder, into a disordered state in which its molecules are distributed at random throughout the room; but the molecules move from the cylinder into the room because of their kinetic energy of movement, and it is the automatic dispersal of that energy which makes the gas escape.

There are various ways of stating any physical law, so we should use the one which is easiest to understand and most clearly expresses the law's driving force and its effects (or perhaps use whichever definition is most suited to each individual circumstance). The easiest and most useful way of stating the second law of thermodynamics, when trying to comprehend the chemistry of life, is to say that 'energy tends to disperse towards an even distribution', or, as P. W. Atkins has put it in his superb book, *The Second Law*, the second law of thermodynamics identifies 'the chaotic dispersal of energy as the purposeless motivation of change . . .'.

You might be able to think of various instances where the law might seem to be violated, for example when the interior of a fridge becomes colder by transferring heat from its already cool compartment into the warmer air all around. Such apparent violation of the second law, however, must be powered by a greater and compensating dispersal of energy elsewhere, so that overall the energy of the universe has become more dispersed. The second law of thermodynamics merely requires that in any change energy will, *overall*, have moved towards a more even distribution. It permits small *local* violations provided they are more than compensated for by energy dispersal elsewhere. Such apparent violations of the principle of energy dispersal will be discussed again later, but for the moment accept that any change anywhere in the universe is accompanied by an overall dispersal of energy from regions of high energy to regions of low energy. How does that help to explain why chemicals should participate in chemical reactions?

All the atoms and molecules and ions of this world are constantly rushing about, colliding with and bouncing off one another. When two chemicals collide, their original neat and balanced electronic structures can be thrown into temporary chaos. Electrons which were previously orbiting quite 'happily' around the nucleus of some atom will suddenly experience new pushes and pulls from the electrons and nuclei of the chemicals they have bumped into. The energy of the collision will force electrons together into places where the electro-magnetic force would not normally permit them to be, and it can force the nuclei of interacting atoms into uncomfortable proximity as well. At the height of the collision the previously stable world of the interacting chemicals becomes a high energy chaos, with the electro-magnetic force straining to restore some sort of stability and balance (see figure 2.7).

One way for stability and balance to be restored is for the chemicals

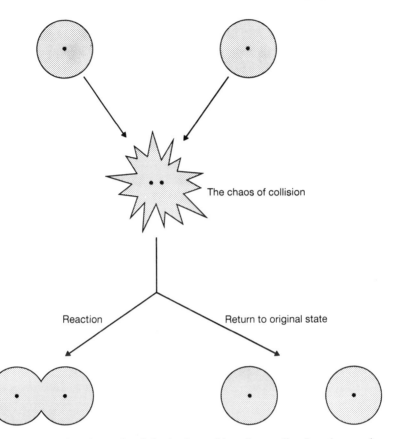

Figure 2.7   When chemicals collide, the chaos of the collision offers their electrons the opportunity to regroup into new arrangements corresponding to successful chemical reactions, or to bounce back to their original states.

to fly apart. If they literally bounce off one another then they will return to their former stable states and no reaction will have occurred. An alternative, however, is that the chaos of the collision might allow the electrons to settle down into the new arrangement of a chemical (or chemicals) composed of parts of both reacting chemicals joined together in some way. Now in many cases the new arrangement might be an inherently *lower energy* arrangement than the arrangement of the original chemicals involved. In the new configuration, in other words, the electromagnetic forces of attraction and repulsion between all the electrons and nuclei involved might be more fully satisfied, or less 'strained' than they were before the reaction took place. In such cases the excess energy will be left over as heat energy stored in the motion and vibrations of the chemicals involved; and if the surroundings are at a relatively low temperature this heat energy will soon disperse away from the site of the reaction to be shared out amongst the other particles all around.

Once this energy has dispersed away, it is unlikely to return spontaneously. The reacting chemicals will have settled down into a new and lower energy configuration – a chemical reaction will have taken place.

So, provided the surroundings are in a relatively low energy state, chemical reactions which leave the reacting chemicals in lower energy states than their previous states will be driven forward by the dispersal of energy into the surroundings at large. In the chaos of collisions all sorts of arrangements of the interacting chemicals will arise, but the only ones which will be consistently retained are those which leave the chemicals 'trapped' in lower energy states – trapped by the dispersal into the environment of the energy that would be needed to energize them back to the states they were originally in.

When two hydrogen atoms collide and react to form a molecule of hydrogen, they release heat energy which disperses away into the surroundings and is therefore not available to split the molecule back into free atoms. Similarly, when two oxygen atoms collide to form a molecule of oxygen, the energy difference between the free atomic and molecular states is dispersed away as heat. If you mix some hydrogen and oxygen molecules and drop a match into the mixture (to provide some extra heat to speed up a few of the molecules) an explosive reaction between hydrogen and oxygen will ensue. In this case the effect of the dispersal of heat energy away from the reaction is obvious – it might kill you if you are standing too close; and at the end of the reaction the shattered walls of the reaction vessel will be covered in droplets of water produced when the hydrogen and oxygen molecules rearranged themselves into molecules of water.

The assistance of the match was needed because at room temperature the molecules of hydrogen and oxygen would not be moving fast enough to collide with sufficient force to allow the water molecules to form. The electron rearrangement needed to make water molecules from hydrogen and oxygen is so drastic that the reacting molecules must collide with greater force than they do at room temperature. Many chemical reactions need some added 'help' like this to make them go. The extra energy supplied by a match or a bunsen burner or whatever raises the energy of the reacting chemicals above a threshold known as the 'activation energy' of the reaction concerned, providing them with the energy needed to participate in a successful reaction. Once a few molecules of water had been formed in our example, the energy released during their formation would be sufficient to jolt many more molecules of hydrogen and oxygen into reacting. Thus, after a little initial 'encouragement', the reaction would proceed along its self-fuelling and explosive way.

That is nearly the whole story of chemical reactions, or at least the rock bottom basics of the whole story, but not quite. Suppose the immediate surroundings of some chemicals contain a great deal of energy indeed, or suppose energy is being directed into some chemical mixture in the form of electromagnetic radiation or in some other way. In such cases the automatic dispersal of energy will make it disperse *into* the chemical sample, rather than away from it. This energy can jolt the chemicals into new arrangements of *higher* energy than the arrangements they were orginally in. So chemistry is not a simple case of atoms reacting to end up in lower energy configurations. That is often what happens, because the surroundings around chemical reactions are often in relatively low energy states. But if the dispersal of energy takes it *into* chemical systems from higher energy surroundings, then chemicals can react to end up in much higher energy configurations.

I can only explore the most basic fundamental principles of chemistry here, not all the details. In many, perhaps even most, chemical reactions, the situation is not quite as simple as I have implied. In real chemical reactions the various energy transactions are often rather complex and varied, involving a combination of different processes, some of which may involve energy moving towards a more even distribution, while others involve energy moving away from an even distribution. The net effect of all the various energy changes is what really matters, however, and in all cases that net effect brings about an increase in the dispersal of the energy of the universe towards a more even distribution.

So what is the fundamental driving force of all chemical change? The answer is the automatic and inevitable tendency of energy to disperse towards an even distribution. If that tendency will take energy *away* from reacting chemicals, then they will react to end up in new configurations of lower energy than their original states; but if it takes energy into reacting chemicals, then they will react to end up in higher energy states. Chemical reactions are the result of chemicals moving towards energy states that are compatible with the energy of their surroundings.

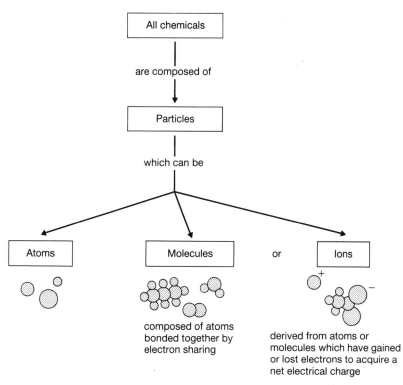

Figure 2.8   A summary of chemistry.

Living things are surrounded by the relatively low energy world of a small planet orbiting cold, dark space far out from the central high energy furnace of the sun. This means that many of the reactions which occur in living things take chemicals down into lower energy states than they were previously in. As living things live, their chemical reactions produce heat energy which disperses out into the colder world around them.

Some absolutely vital reactions of life, however, involve chemicals being pushed up into higher energy states by the electromagnetic radiation they receive from the sun. The basic chemical interaction which powers almost all forms of life is the absorption of the energy of sunlight by molecules such as the 'chlorophyll' in the leaves of green plants. This raises the chlorophyll molecules into a high energy state, and this energy is then trapped by a series of other chemical reactions and stored in the chemical substance of living things.

The molecular mixing bowl

In a world full of energy nothing is stationary. The seemingly calm solidity of the matter all around you, and within you, belies the perpetual seething internal motion which exists in the microworld of molecules and atoms and ions. All molecules and atoms and ions are moving. Many are darting about and bouncing off one another in a constant chaotic random dance. This is especially true in gases and liquids, but even in solids the constituent molecules and atoms and ions are constantly jostling against one another and often vibrating internally like tiny sub-microscopic springs. All chemical bonds behave a bit like tiny springs, constantly being stretched and compressed as the chemicals they are part of are jostled about by the motion of the other chemicals all around them.

All this internal activity is known as the 'thermal motion' of the microworld, since it is caused by the fact that all molecules, atoms and ions possess a certain amount of the energy of motion we call heat. The influence of thermal motion on the chemistry of the world is vital, so vital that without it chemical reactions would not take place. For a chemical reaction to occur, two or more chemicals must bump into one another, and they are never going to bump into one another if they are not moving. Thermal motion turns much of the chemical microworld into a 'molecular mixing bowl'. The mixing action is most effective when the chemicals involved are in solution, since the relatively loose structure of a solution allows all the dissolved chemicals, and all the molecules of the solvent, constantly to be mixed together thoroughly and at random. Much of the chemistry of life takes place in solution, although in a much more structured and less chaotic solution than a simple flask of water. The water of the cell cytosol surrounds all the chemicals of the cell and permeates into all available clefts and pores and crevices of the molecules the cell contains. So the solution of the cytosol acts as an all pervading chemical sea in which many of the chemicals of life are mixed together by random thermal motion as if in a molecular mixing bowl.

The chemical reactions of life are often very complex, involving many different chemicals moving from place to place in the cell and reacting together in specific sequences. When we first examine some of the complex chemical events within the cell, and see, for example, chemical A meeting up with B to form C, which moves elsewhere to react with D to form E, which splits into F and G which move apart and each enters another sequence of reactions . . ., it is sometimes hard to believe that *random* thermal motion can bring about such seemingly purposeful change. This is a fundamental concept which we should examine further – the ability of random thermal motion to generate directed and seemingly purposeful change.

To appreciate the power of random motion to bring about seemingly purposeful change, imagine a room full of blindfolded people all instructed to walk about at random 'bouncing' off the walls and one another. Imagine also that they have been told to stop moving only when they bump into a small picture hanging from a wall. Finally, suppose that all the pictures are hung in a second room, linked to the room full of people by a narrow open doorway (see figure 2.9). What would happen?

They would all start moving and bumping into one another and the walls of the first room in a random 'thermal' chaos which, at first, would seem to be getting them nowhere. Eventually, however, someone would inevitably end up walking towards the doorway and so

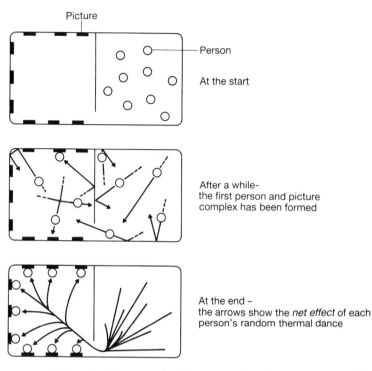

Figure 2.9   The molecular mixing bowl party game. Random movement combined with specific interactions between specific objects when they meet can generate seemingly purposeful change.

would walk through into the next room. In the next room they would continue their random walk until eventually, probably quite soon, they would encounter one of the pictures and stand still by it. In the meantime, some of the others would have found the doorway, purely by chance, and they too would stumble about in the second room until encountering a picture. Gradually the chaos would clarify and order begin to emerge. Inevitably, everyone would eventually make it into the room with the pictures. A few might move back into the first room, of course, but there are no pictures in there so their random walk would eventually take them through the doorway once more. As more of the pictures were found, it would become increasingly unlikely for the last few people to find one, but they had been instructed to continue walking until they did find one. The last couple of people might make several unsuccessful sorties into the room with the pictures, then back into the first room, then back again, but they would eventually meet up with the unoccupied pictures, assuming there was a picture available for everyone. The simple but highly effective chemical party game would be over.

Anyone observing only the start and the finish of the game – its overall effects, in other words – would think that the movement of the players had been very purposeful indeed. They would surmise that each person moved from the first room into the second room to find and stand still by a picture. They would not realize that there was no direction or purposefulness behind this movement. It was brought about simply by the random motion of each individual, combined with the rule that when a picture was discovered the motion should stop. The combination of random motion and simple specific inter-actions between certain objects when they meet can be a very effective way to get things done in a seemingly purposeful manner.

All chemical reactions, including the many complex reactions and sequences of reactions of living things, proceed thanks to the same sort of random dance which is given seeming purpose and direction only by the reactions and interactions of chemicals once they eventually meet. Thus, the statement 'A travels out from the nucleus of the cell and reacts with B, to form C, which travels to a specific location X, at the cell membrane, and reacts with D which is bound to X to form E', seems to imply that the chemicals concerned somehow know where they have to go and what they have to do. They do not. What might happen is that A would bump about at random within the nucleus until it encountered a pore in the nuclear membrane through which it would pass, provided its random trajectory was taking it in the right direction. A would then bump about in the cytosol until, by chance, it encountered B, with which it would automatically react to form C, which would not then 'travel' purposely towards site X on the membrane, but would bump about at random until it happened to hit site X and then react with the D which it encountered there, forming E. Large numbers of molecules of A in the nucleus could eventually follow this route, to form a large pool of E beside the cell membrane in the region of X. Once again a seemingly purposeful series of movements and events would have been brought about by random motion and the automatic reaction between chemicals which can react together when they encounter one another.

The interior of a living cell is certainly far more structured and complex than a simple mixing bowl. Some of the chemicals which give the cell structure can serve to assist the movement of other chemicals from place to place, by forming channels through which the chemicals can pass, for example, or by forming barriers which prevent them from moving into certain regions, or by actually binding to various chemicals to assist them on their way; but all interactions between chemicals of the cell rely on random thermal motion to bring them about. Certainly, various chemicals can have groups of atoms exposed on their surfaces which allow them to stick to one another, or which perhaps allow them to be drawn together by electromagnetic attraction once random motion has brought them sufficiently close; but at heart the living cell is a chemical mixing bowl in which the mixing is done by the random agency of thermal motion.

This constant motion can serve to break things apart as well as bring them together. All the chemical components of the cell must be able to survive their constant thermal jostling, or else be broken up by its effects. In many cases the motion serves to break up the complexes which form between chemicals once they have reacted, causing free products to be released. Chemicals A and B, for example, may join together into the complex AB, and then participate in a reaction which changes AB into a new complex CD. Once CD is formed, the rigours of random thermal motion may cause the weak bonds holding it together to fall apart, releasing C and D as the final products of the reaction. It is also possible that random thermal motion may cause the complex AB to fall apart before the reaction has had a chance to proceed, releasing free A and B to bump about some more until they meet up once again in a more succesful liaison. Equally, thermal motion might bring C and D together to *reform* CD, and sometimes some of that CD might undergo the reverse reaction back into AB and then A plus B. Provided more C and D is being formed *overall* than A and B, then, overall, A and B will be seen to be reacting to generate C and D. That seemingly straightforward and unidirectional chemical reaction, however, could be the tidy net effect of a messy bidirectional chemical chaos.

Obviously, chemicals such as A and B do not 'know' that they can react together, the reaction simply occurs automatically when they meet. Equally, A does not know that it cannot successfuly react with other chemicals such as X and Y. It may be possible (it often is) for a chemical such as A to form weak complexes with many different chemicals such as X and Y, but if no proper reaction can take place within the complexes AX and AY they will soon be broken apart by thermal jostling, making the A available again for B to meet up with it and react with it.

The overall chemical behaviour of a living cell, or any other collection of chemicals, is the end product of a myriad of different random encounters and interactions between the chemicals involved. Many of these encounters will result in no chemical reaction, because the chemicals concerned simply bounce off one another, or form short-lived complexes held together by weak bonds which are soon broken, or bump into one another in the wrong orientation, or with insufficient energy, and so on; but a few of the encounters will allow chemicals which can react with one another to collide with sufficient force to initiate the appropriate reaction, and in the particular relative orientation which allows the reaction to proceed. Chemistry at the level of individual atoms and molecules and ions is largely a chaotic mess, with the unfruitful encounters often far exceeding the en-counters which result in successful reaction; but viewed from our lofty height, which usually only allows us to observe the net effects of an unseen chaos, chemistry can appear to be smoothly converting a specific set of reactants into a specific set of products.

The frantic chaos of chemistry proceeds too fast and too remotely for us to follow it without great difficulty. We are in the position of airborne observers who see trainloads of shoppers flowing into a city on Christmas Eve morning, and trainloads of the same shoppers laden with purchases flowing back to the suburbs in the evening. From the air we can see the overall effect of suburban shoppers 'reacting' with the shops full of goods, but we remain unaware of the hidden random chaos which allows the reaction to proceed!

So, throughout this book, when you come across such phrases as 'A moves from site X to site Y and reacts with B and C to form D, which then moves into site Z to react with E and form F' remember that all this seemingly purposeful motion and reactivity proceeds thanks to the random action of the molecular mixing bowl, with thermal motion moving things around until they just happen to meet up in a manner which allows them to react as required. In every case it might be more correct for me to say such things as 'A bumps about a bit until it hits J, which it can't react with, so it bounces off and hits K, which it forms a brief association with until the complex is split up and the free A bumps about a bit until it happens to fly through the passageway connecting site X to site Y, where it bumps about a bit until it hits L, which it bounces off and into M, which it bounces off and hits O, which it bounces off and hits the complex BC, but without sufficient energy to react, so it bounces away to hit a few more chemicals until it happens to hit another BC complex with sufficient energy for the reaction to proceed which forms D, which bounces about until it happens to move from site Y to site Z, where it bounces about a bit until it meets E and reacts to form a complex DE, which unfortunately is hit so hard by a fast-moving neighbour that it splits up into D and E again, allowing D to bump into N and form a brief association with it until the complex DN is split up by thermal motion to allow D to bump into O, then P, then a molecule of E to form the complex DE which reacts to give F!'

The book would get rather tedious if I kept that up, and in any case the details would be different for every molecule of A that reacted, and for every molecule of all the thousands of other chemicals undergoing other reactions. Far better to say simply 'A moves from site X to site Y and reacts with B and C to form D, which then moves into site Z to react with E and form F', since that is what happens overall, if we give the random molecular mixing bowl time to sort it all out.

Our exploration of the world of fundamental physics and chemistry is over. The one fundamental phenomenon not yet mentioned is that of 'space-time', which is simply the place where all the things talked about so far are to be found, and where all the changes and interactions happen. The three dimensions of space can be united with the one dimension of time to create the four-dimensional world of space-time in which matter and charges and forces and energy play out their eternal dance. It is time to see how some of the steps and flourishes of that dance give rise to life.

3 Proteins make genes make proteins

To a biochemist, life is the overall result of a network of chemical reactions – many thousands of chemical reactions which, taken individually, are the simple results of chemicals bumping into one another to cause rearrangements of their electrons, but whose net effect is a living thing of awesome complexity.

Biochemists will admit to no mysterious aphysical or spiritual 'vital principle' needed to turn a 'bag' of reacting chemicals into a living cell. A few might baulk at including human consciousness, thought and our apparent 'free will' in the list of things which 'mere chemicals' reacting together can do (especially those who hold some religious faith) but none will dispute that the physical activities of his or her body are the result of the physical interactions of the chemicals within.

How can we be so confident that chemistry can automatically construct an oak tree from an acorn, or a professor from a fertilized human egg?

In chapter 1 I summarized the events which biologists believe make up the essential central mechanism of life. They believe that genes embodied in molecules of DNA can cause the production of molecules of RNA, which can cause the production of molecules of protein, which can catalyse all the chemical reactions of life, including the ones involved in allowing DNA and RNA to cause proteins to be manufactured. We must now take a look at the chemicals involved in this central mechanism, to see how they manage to work such apparent wonders.

Proteins - sculptors of the cell

Proteins are the 'molecular workers' that actually construct and maintain all cells. That is what I told you in chapter 1, but it is worth reflecting a while on the significance of that bold assertion. It claims that all the astonishing variety and complexity of life on earth is a result of the chemical activities of just one particular class of chemical compound - the proteins. Other types of chemicals are involved indirectly, as the genetic materials that specify which proteins there should be, for example, or as the raw materials that the proteins act upon, or as the products the proteins create; but life is essentially a work of chemical art sculpted by proteins.

If that is the case, then the proteins must obviously be a most remarkable and versatile class of compound. Proteins certainly are both remarkable and versatile, but in many ways they are also surprisingly simple. The structures and activities of the proteins provide further examples of great variety and complexity arising from simplicities within.

All proteins begin life as long chain-like molecules made when many smaller types of molecules - the 'amino acids' - become linked together (see figure 3.1). It is at the level of amino acids that the simplicity is found, because only 20 different types of amino acid are used as the 'building blocks' of all proteins. The average protein contains about 350 amino acids linked together. Many thousands of different proteins exist in nature and a virtually infinite number of different proteins *could* exist; but all must be composed of different combinations of the same set of 20 available amino acids.

These 20 amino acids comprise the chemical 'alphabet' from which the story of protein-based life (i.e. all life on earth) is constructed.

Most proteins do not stay as long chain-like molecules for long. Instead, they tend to fold up. Some proteins fold up into compact 'globular' shapes. Others become coiled into single or multiple helices. Some become bound by weak forces of electromagnetic attraction to other protein molecules, to form 'multi-subunit' aggregates (see figure 3.2). It is in their final folded form that proteins can actually perform their various roles as the sculptors of the cell, and the way in which any

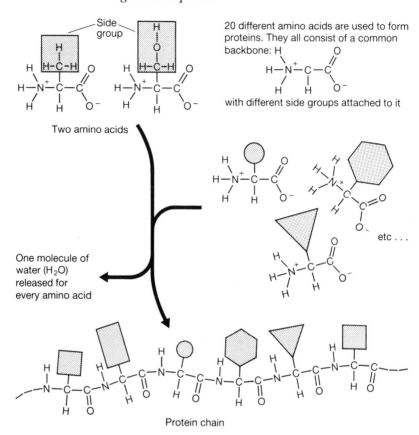

Figure 3.1　*Proteins are composed of long chains of linked amino acids. Twenty different types of amino acids are found in proteins, and most proteins are hundreds of amino acids long.*

individual protein will become folded is determined very precisely by the amino acids it contains and the sequence in which they are linked together.

Why should proteins fold up into specific three-dimensional conformations and what exactly makes them do so? The short answer is that protein folding is controlled by chemical interactions between various atoms of a protein, and between the protein's atoms and the water molecules all around, but we should consider the mechanisms of folding in a little more detail.

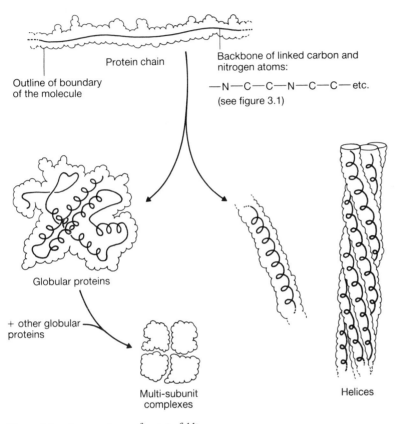

Outline of boundary
of the molecule

Protein chain

Backbone of linked carbon and
nitrogen atoms:

—N—C—C—N—C—C— etc.
(see figure 3.1)

Globular proteins

+ other globular
proteins

Multi-subunit
complexes

Helices

Figure 3.2   Some patterns of protein folding.

The surface of any chemical is able to interact with other chemicals in a variety of ways. First, various atoms of the chemical may participate in full chemical reactions with another chemical, to result in the two chemicals becoming linked by covalent or ionic bonds. Such chemical reactions have some relevance to protein structure, because the atoms of some specific amino acids can form covalent bonds with the atoms of other amino acids *of the same protein* to form covalent 'cross-bridges' which hold two distant parts of a protein molecule together. These cross-bridges usually form, however, *after* a protein has folded up in such a way as to bring the necessary amino acids together, so they tend to stabilize a protein's final conformation rather than create it in the first place. Ionic bonding is also important to the maintenance of protein conformation. Various amino acids carry full positive charges while others carry negative charges. Obviously, the force of electromagnetic attraction between these opposite charges can hold various parts of a protein together, while forces of electromagnetic repulsion between amino acids carrying the same type of charge can keep various parts of a protein apart.

So covalent and ionic bonds can be vitally important in determining a protein's folded structure, but the most influential forces involved in protein folding do not involve full covalent or ionic bonds. They depend instead on rather weak electromagnetic interactions of a more subtle nature.

Many of the chemical bonds which join up the atoms of a protein chain are polar covalent bonds. In other words, the bonds are 'polarized' into regions with a slight positive charge (δ^+) and regions with a slight negative charge (δ^-). So the surface of a protein can be covered in regions of slight positive and negative charge. The effects of these weak charges on protein folding are similar to the effects of full charges, though weaker, causing oppositely charged regions of the protein to be drawn together while keeping similarly charged regions apart.

The most important single influence on the folding of a protein, however, is probably the interaction between the atoms of the protein and the molecules of water surrounding it. To understand this influence you must know a little about the interactions between water molecules themselves. Water molecules contain two polar covalent bonds (see figure 2.5) causing the two hydrogen atoms of a water molecule to carry δ^+ charges while the oxygen atom carries a δ^- charge. This causes the hydrogen atoms of one water molecule to be attracted to the oxygen atoms of others, and vice versa, so an extended network of weak electromagnetic attractions tends to hold different water molecules together. So there is a loose internal structure within any region of water, caused by the water molecules being electromagnetically attracted to one another.

Some 'hydrophilic' (water-loving) amino acids have a great chemical affinity for water. These are generally the amino acids which carry full or partial electric charges which can readily become integrated into the network of interacting charges between the water molecules themselves. So amino acids which can interact electrically with water molecules can slip into any body of water relatively easily.

Other amino acids, however, have no affinity for water at all. These are generally the amino acids which carry no regions of positive or negative charge, and which are therefore unable to interact with the partial charges on water molecules. Regions of a protein carrying such 'hydrophobic' (water-hating) amino acids experience a force pushing them together, away from the water molecules, into a tightly bundled 'core' of the protein. The origin of this force is really the force of attraction between water molecules, since this tends to 'squeeze' the hydrophobic amino acids out of the water structure as much as is possible. So the electromagnetic force pulls water molecules and those amino acids which can interact with water together, and in so doing it squeezes hydrophobic amino acids away into a small volume which disrupts the structure of the water as little as possible.

Once hydrophobic amino acids have become squeezed into the central water-free core of a protein, they are themselves partly held together by weak forces of electromagnetic attraction known as 'Van der Waals forces'.

To understand the origin of Van der Waals forces you must appreciate that, on the surface of any chemical, regions of very slight positive and negative charge are constantly being formed and then destroyed, due to the continually changing distribution of the electrons. It is as if the random orbiting of all the electrons continually creates and then destroys regions of partial electric charge where electrons have fleetingly become concentrated in one region and diminished elsewhere. These regions of transient electric charge come and go constantly on the surface of all chemicals, and they create an electromagnetic force which weakly attracts all chemicals to all others. This attraction occurs when a region of transient positive charge on one chemical finds itself opposite a region of transient negative charge on another. The two chemicals will obviously be drawn together, and in being drawn together they will cause their mutual attraction to be reinforced, since the positive charge on one chemical will draw electrons of the other chemical towards it, and so reinforce the other chemical's negative charge, while that negative charge will repel the electrons of the first chemical and so reinforce its positive charge.

Of course it is equally likely that two regions of the *same* charge on two chemicals might end up next to one another, but such repulsive interactions will not reinforce themselves, they will tend to destroy themselves. A fleeting negative charge on one chemical, for example, would repel electrons from a neighbouring region of fleeting negative charge on another chemical, and so encourage that charge to disappear. Equally, a region of fleeting positive charge on one chemical would attract the electrons of any other chemical and so tend to destroy any adjacent region of positive charge on that chemical.

So the surface of all chemicals is 'alive' with constantly fluctuating small electrical charges. When two opposite charges of this type arise on two neighbouring chemicals they tend to draw the chemicals together due to a self-reinforcing electromagnetic attraction; while regions of the same type of charge on neighbouring chemicals tend to destroy one another. The overall effect of these transient small charges, provided they are not swamped by other more powerful forces, is to draw neighbouring chemicals together to form weak links between them known as Van der Waals 'bonds' (or Van der Waals forces). The hydrophobic amino acids which are pushed together into the core of a folding protein are held together by weak Van der Waals bonding.

The forces which determine the way in which a protein will fold up are all essentially electromagnetic forces of attraction and repulsion operating in various ways and situations. The electromagnetic forces of attraction between water molecules, and between water molecules and hydrophilic amino acids, tend to force hydrophobic amino acids into the centre of folding proteins where they will cause the minimum disruption to the external network of electromagnetic attractions. The hydrophobic amino acids will be held together by weak and fluctuating Van der Waals forces of transient electromagnetic attraction. Regions of a protein chain carrying opposite permanent charges will be attracted to one another, while regions carrying like permanent charges will be repelled; and in both cases the charges concerned may either be the full ionic charges caused by the loss or gain of electrons, or the partial charges associated with polar covalent bonds in which the electrons are shared unequally between the atoms which they bond together.

So any newly formed protein is soon pushed and pulled into whichever folded conformation allows all the various electromagnetic attractions and repulsions to be most satisfied. Attractions and repulsions between the protein and surrounding water molecules are probably the most important, but are certainly not the only ones that matter. There are also various attractions and repulsions between the different parts of the protein chain itself; but they all resolve themselves into the most stable final folded form for the protein concerned. The electromagnetic force forces proteins into very specific folded conformations which are dependent only on the amino acid sequences of the proteins concerned.

The previous sentence summarized the second great simplicity underlying the complexity of the proteins: there is a virtual infinity of different ways in which protein chains could become folded into different complex shapes, but in each case the final structure of any protein is determined only by its amino acid sequence. In other words, it is determined by which of the set of 20 amino acids are found in the protein chain, and the sequence in which they are arranged. Linear chains assembled from only 20 different amino acids can generate a virtual infinity of folded proteins, all with different and very specific three-dimensional structures; and these specific three-dimensional structures determine what the proteins can do. And remember, once proteins are formed (we will see how they are formed later) they fold up automatically, thanks to the pushing and the pulling of the electromagnetic force.

In chapter 1 you were told of some of the things that protein molecules can do – that they can act as enzymes, speeding up particular chemical reactions, or that they can act as a form of solid and secure scaffolding within the cell, or as muscle fibres, or transport proteins that carry other chemicals from place to place, or as chemical messengers that flit from cell to cell to control and co-ordinate a variety of cellular activities, and so on. . . .

The full repertoire of the powers of proteins will be explored in chapter 4, but in the meantime we should consider in general terms how these powers come about. One short and simple answer is as follows – when a protein adopts, automatically, its final folded form, it presents a very specific chemical surface to the environment around it.

The precise shape and chemical nature of this surface determines what the protein will be able to do.

The surface geometry of an enzyme molecule, for example, will contain clefts and grooves into which only the chemicals with which the enzyme must interact can fit. It will also contain sites which can bind to any chemicals which serve as 'cofactors' which help the enzyme to perform its task. Once any necessary cofactors are bound, the surface of an enzyme or enzyme/cofactor complex will contain a particular array of atoms and ions which are electromagnetically attracted to, or may even form strong covalent bonds with, the chemicals involved in the reaction which the enzyme catalyses; and it will contain a particular array of atoms and ions which happens to encourage the reaction catalysed by the enzyme to proceed. The surface of an enzyme catalyses a specific chemical reaction because the precise geometry and chemical nature of that surface allows the electromagnetic force to push and pull at the electrons of chemicals which the enzyme can bind to. This pushing and pulling is what directly encourages the rearrangement of electrons which constitutes a chemical reaction.

So each enzyme presents to the environment a very specific 'active surface' whose geometry and chemical nature allow it to bind to specific chemicals and catalyse a specific reaction between them; and the chemical groups which do all this binding and catalysis are the atoms and ions of the enzyme's amino acids, often assisted by the atoms and ions of chemical cofactors which the enzyme can become bound to.

Similar logic applies to the activities of all proteins. The surface of a transport protein, for example, will contain an array of atoms and ions which can bind to the chemical to be transported in some conditions, and release that chemical in other conditions. The surface of a protein that acts as a chemical message, such as a 'hormone', will contain an array of atoms that allows it to bind to some specific part of certain cells and so alter those cells' chemical activities in some appropriate way.

A protein folds up into a conformation which is determined by its amino acid sequence, and which presents to the environment around it a chemical surface which allows the protein to perform its particular chemical task; and the folding and the performance of the task (and, indeed, the creation of the protein in the first place) all proceed automatically governed only by the physical laws and forces of nature – particularly the electromagnetic force.

Proteins are huge and rather complex molecules, composed of thousands of atoms; but the activities of proteins are governed by the same fundamental physical and chemical principles as we saw operating on the small and simple molecules used as examples in chapter 2. Proteins are masssive, magnificent and very powerful, but there is no mystique surrounding their many varied activities. They are chemicals whose 'every move', like the 'moves' of all chemicals, is the result of a few fundamental and rather simple principles operating in a wide variety of complex situations.

I have already mentioned that proteins sometimes become firmly attached to other small molecules or ions which act as cofactors which assist a protein in its task. These cofactors sometimes become virtually integral parts of the completed proteins, while in other cases they form much looser and more readily reversible associations. Many of the vitamins and minerals which are needed in our diet are needed because they form the cofactors required by various enzymes and other proteins.

I have also mentioned that very often two, or three, or more individual protein molecules can become bound to one another (by weak forces of electromagnetic attraction), to form large multi-subunit complexes able to perform specific tasks. Another complication is that some proteins become permanently chemically modified, by having various small chemical groups covalently attached to them at specific sites, and only then are they able to perform their appropriate tasks (see figure 3.3). The modification processes, of course, are catalysed by other protein molecules.

One of the most common forms of protein modification involves the covalent bonding of various 'sugar' groups to the protein surface. All sugars belong to the class of chemical compounds known as 'carbohydrates', which are largely composed of the atoms carbon, hydrogen and oxygen. Proteins which have sugar groups attached to their surface are known as 'glycoproteins' and in many glycoproteins the sugar groups must play vital roles in influencing the molecules' overall chemical activities. What precise roles are played by the sugar groups of glycoproteins is often rather unclear, and is the subject of an

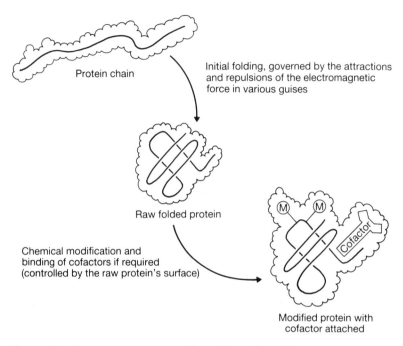

Protein chain

Initial folding, governed by the attractions and repulsions of the electromagnetic force in various guises

Raw folded protein

Chemical modification and binding of cofactors if required (controlled by the raw protein's surface)

Modified protein with cofactor attached

Figure 3.3   Some proteins must undergo chemical modification and/or become attached to various chemical cofactors before they attain their mature active form.

active and expanding field of modern research. Many of the most‘ important glycoproteins are found protruding from the surface of cells, and it seems that their sugar groups may be crucially involved in the specific chemical interactions between different cells, and between cells and the various molecules floating free in the extracellular environment.

When I refer throughout this book to 'proteins', you should bear in mind that usually what I really mean is 'proteins, or glycoproteins, or proteins modified in some other way'. It would be tedious to state repeatedly whether I am referring to a naked or 'raw' protein, a glycoprotein, a protein which has had various phosphate groups attached to it, and so on; or to remind you constantly that raw proteins composed of linked amino acids alone are often modified by the addition of various small chemical groups. I use the term proteins to refer to both raw unmodified proteins, and those which have undergone subsequent modification.

So we should remember that the chemical activities of a protein may not all be produced solely and directly by the atoms belonging to the proteins' amino acids. The atoms belonging to sugar groups, and the various other chemical groups which can be added on to proteins, can also have crucial roles to play. All such modifications are either catalysed by other proteins, or occur spontaneously; and they occur because the structure of the raw protein, as determined by its amino acid sequence, makes them occur. So, despite the complications introduced by the process of protein modification, the prime importance of a protein's amino acid sequence as the determinant of its chemical activity is maintained.

The common feature uniting the activities of all proteins, raw or modified, is that they depend on selective binding to other chemicals to allow them to achieve their feats of highly selective catalysis, transport, structural organization, or whatever. The force which makes this selective binding possible is, once again, the electromagnetic force. It may act in the guise of Van der Waals forces, or of electromagnetic attractions between the partially charged atoms involved in polar covalent bonding, or of electromagnetic attractions between fully charged ionic groups, or of full covalent bonds formed when electrons become shared between a protein and the chemicals it interacts with; but in all cases it is the electromagnetic force which allows proteins to interact selectively with specific chemicals and make significant things happen.

We have now explored in fuller detail two of the three secrets of life revealed in the Introduction. We have seen how 'the electromagnetic force interacts with the energy of the world to make chemical reactions proceed'; and we have examined a few fundamental principles concerning the ability of proteins to 'directly or indirectly promote the specific chemical reactions needed to create and sustain all life'. We must now explore the meaning of the other secret, which tells us that 'chemicals called nucleic acids are able to form and direct their own reproduction and also to direct the production of chemicals called proteins.'

DNA and RNA - the genetic moulds

Take a look at figure 3.4a and you will see the nucleic acid called DNA in its full molecular glory. Each sphere represents one atom of this massive and magnificent structure, and five different types of atom are present - carbon, hydrogen, oxygen, nitrogen and phosphorus. This is the structure which transmits genetic information from one generation to the next. Subtle variations in its basic plan determine whether our eyes are blue or brown or green; whether our hair is black or blonde or brown or red; whether we have four limbs, or six, or a hundred; whether we have backbones or exoskeletons; whether we consist of trillions of cells or only one, whether we can run amongst the bushes or are one of the bushes.

If you look carefully at figure 3.4a you can see why DNA is known as a double-helix. There are two helical ribbons of linked atoms winding around a central core. A section of one of these helical ribbons is highlighted in the figure by lines drawn on either side of it. Of course you cannot see all the atoms of the double-helix in the figure, because it is a cylindrical three-dimensional object. To appreciate its structure fully you would need to walk around a model, but if the double-helix in the figure were slowly rotated the view would essentially remain unchanged.

Each of your cells contains an enormous amount of this DNA, altogether many millions of times as long as the short section shown in the figure. How do these lengthy 'threads of life' have such an influence over our existence? To answer that question, we will need to examine the structure of DNA more closely, and simplify it until we uncover the essential mechanisms by which it works.

In figure 3.4b(i) all the individual atoms have been merged into a unified molecular structure with only the molecular boundaries drawn in. Some slight artistic licence has been used to illustrate one of the most important aspects of the double-helix - it is actually composed of two separate single-stranded DNA molecules wound around one another. These two strands of the double-helix are held together by weak forces of electromagnetic attraction. Obviously, weak Van der Waals forces could be expected to operate along the interior of the

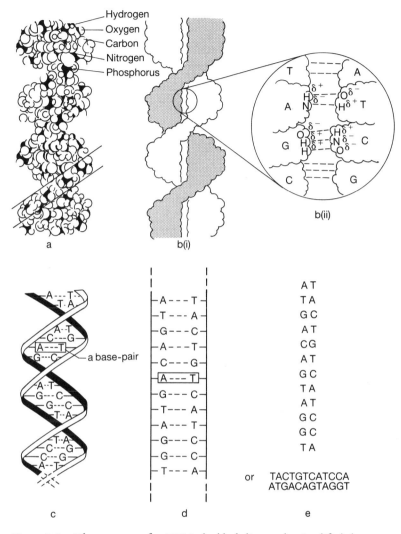

Figure 3.4   The structure of a DNA double-helix can be simplified down to a sequence of letters representing the bases A, T, G and C.

double-helix, where the two strands nestle against one another, but these are much less important than the weak bonds illustrated in the magnified illustration of figure 3.4b(ii). The interior, or core, of the double-helix contains only four different types of chemical groupings known as 'bases', whose names are 'adenine' (A), 'thymine' (T), 'guanine' (G) and 'cytosine' (C); and each of these bases carries just two or three atoms that are actually involved in the bonding between the two helices.

The base known as A carries a nitrogen atom with a slight negative charge and a hydrogen with a slight positive charge. The base T carries an oxygen with a slight negative charge and a hydrogen with a slight positive charge. The geometry of these two bases, and the precise way in which they are held by the DNA backbone, allows their respective slight positive and negative charges to lie opposite one another in a way that will hold them together. This causes a structure known as a 'base-pair' to form, in which the bases A and T on opposite strands of the double-helix are held together by weak forces of electromagnetic attraction.

Much the same sort of base-pairing can occur between the other two types of base – G and C. The base G carries two hydrogen atoms with slight positive charges and one oxygen with a slight negative charge. The base C carries one hydrogen with a slight positive charge, plus a nitrogen and an oxygen with slight negative charges. Again the geometry of the two bases and the way in which they are held by the DNA backbone allows their respective positive and negative charges to lie opposite one another in a way that holds the two bases together.

These two types of base-pairs hold the entire length of any DNA double-helix together. If you examined any strand of a double-helix, you would find a particular sequence of the four bases A, T, G and C. If you then examined the other strand, you would find a precisely 'complementary' sequence of bases, in which each A on one strand is opposite a T on the other, each T would therefore be opposite an A; each G would be opposite a C and so each C would have to be opposite a G. So the DNA double-helix is held together by the base-pairs A–T or G–C, and the force which holds these base-pairs together is the force of electromagnetic attraction between complementary regions of slight positive and negative charge. The atoms which carry these slight charges carry them because they are involved in polar covalent bonds, in which electrons are shared unevenly between the covalently bonded atoms concerned.

We can summarize and simplify things by adopting the symbolic representation of figure 3.4c, in which the helical backbone of each

DNA molecule is represented as a twisting ribbon, while the chemical groups called bases, which are attached to the backbone, are represented by their respective initials. We can make things even simpler by untwisting the double helix, so that the important sequence of the central base-pairs is clear to see (figure 3.4d). Finally, to complete our simplification process, we can remove the backbones entirely, to end up with a letter sequence representation which carries all the information we need to know about any DNA double-helix (figure 3.4e). The atoms of the backbone do not really matter, they simply hold each DNA molecule together and they are always the same. What really matters is the sequence of base-pairs, because that is what differs between different double-helices, or between different regions of the same double-helix.

Since the base-pair sequence is the only aspect of the double-helix which can vary, this sequence must somehow carry the 'genetic information' which we know the double-helix contains. So, somehow, the genetic differences between you and me and monkeys and magnolias must depend on differences in the sequences in which the two possible base-pairs are arranged in our genes.

If all organisms contained a genome only ten base-pairs in length, then

AAATTTGGGC
TTTAAACCCG

might make a man,

AAAATTGGGC
TTTTAACCCG

might make a monkey, and

GGCCGGCCAA
CCGGCCGGTT

might make a magnolia.

In fact, organisms contain genomes ranging from thousands to many billions of base-pairs in length, but the principle is the same. Different organisms are different because of differences in the genetic information carried within their DNA; and these differences are differences in the sequence in which the two possible base-pairs are arranged into long sequences on the DNA double-helix.

In case you are intrigued by the precise chemical structure of DNA, a single base-pair of a double-helix (unwound for clarity) is shown in its chemical entirety in figure 3.5. This figure reveals that the helical backbone of DNA is composed of alternating sugar and phosphate groups. It also points out that the weak bonds holding the base-pairs together are known as 'hydrogen bonds', due to the central role of a slightly positively charged hydrogen atom in each one. Hydrogen bonds of one sort or another are vital to many different aspects of the chemistry of life. We have actually met them already. The forces of attraction between water molecules, discussed on page 50, are also hydrogen bonds, since the slightly positively charged hydrogen atoms of water molecules play the central role in drawing the water molecules together. Many of the forces of attraction responsible for folding proteins into precise shapes are also due to hydrogen bonds, in which hydrogen atoms with δ^+ charges are attracted to other atoms with a δ^- charge.

The description I have given of the hydrogen bonds holding the double-helix together is actually slightly oversimplified. I have presented the bond as being entirely due to the force of electromagnetic attraction between the partially charged atoms involved. Although this is largely true, these bonds also have a slight covalent nature, due to a degree of *sharing* of electrons between the atoms involved. This is a minor complication of the sort which occurs repeatedly when the interactions of chemicals are examined in detail. Few things are absolute in nature, and it turns out that virtually all ionic bonds have a slight covalent nature and all covalent bonds have a slight ionic nature. Pure ionic and pure covalent bonding should be regarded as the two extremes of a graded continuum. Such complications need not really concern us, for in all cases it is the electromagnetic force, in one guise or another, which holds chemical bonds together.

You now know all you need to know about the structure of the DNA double-helix. It is composed of two separate single-stranded DNA molecules wound around one another and held together by weak forces of electromagnetic attraction. The structure of different DNA strands can vary due to variations in the sequence of bases which

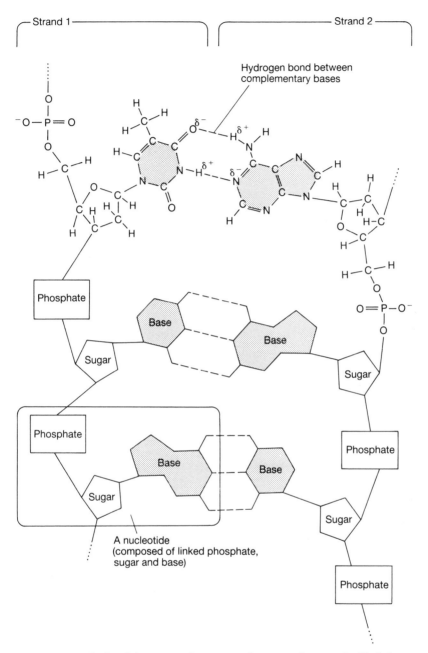

Figure 3.5 The detailed structure of DNA. One base-pair of a DNA double-helix is shown in full. The other two base-pairs, shown in schematic form, reveal that each strand of DNA is composed of linked nucleotides, each nucleotide being composed of a phosphate group, a sugar group and a base (A, T, G or C). The two strands of the double-helix are antiparallel – they run in opposite directions.

are strung out along the DNA backbone; and each of the four bases can form a base-pair with only one of the other bases – A pairing with T, and T with A; G pairing with C, and C with G. So the structure of the double-helix will only be able to form between two *complementary* DNA molecules, in which each base on each strand lies opposite the base it can pair up with on the other strand. The rules of base pairing can never vary, because they are a direct consequence of the three-dimensional structure of the bases themselves and of the molecular backbone which holds them together. The base A is only able to pair with the base T, and the base G is only able to pair with the base C, because these are the only forms of base-pair which allow the double-helical structure to form.

How can this complex and intriguing structure serve as the genetic material which determines how cells and organisms form and reproduce?

To act as the 'genetic material' able to generate endless generations of life, the DNA double-helix must be capable of two main things. It must be able to direct the manufacture of specific proteins – the molecules which actually generate life directly. Secondly, there must be some way for the double-helix to become copied, so that new copies can be produced to be passed on to future generations. DNA 'replication' is the technical term for this copying process, which we will explore first.

The basic principles of how DNA can become copied are very simple to understand, given the rules of base-pairing which state that only the base-pairs A-T and G-C are able to hold the double-helix together. Suppose I pulled apart a model double-helix, like the one shown in figure 3.6, and gave you just one of the separated strands. If I also supplied you with plenty of the four bases, linked to the sugar and phosphate groups which form the DNA backbone, you would easily be able to reconstruct the original double-helix, provided you remembered the rules of base-pairing. You would simply need to construct a new 'complementary' strand, carrying an A opposite each T of the existing strand, a T opposite each A, a G opposite each C and a C opposite each G.

This is very close to what actually happens in nature (see figure 3.7). The building blocks of new DNA are chemicals known as 'nucleoside

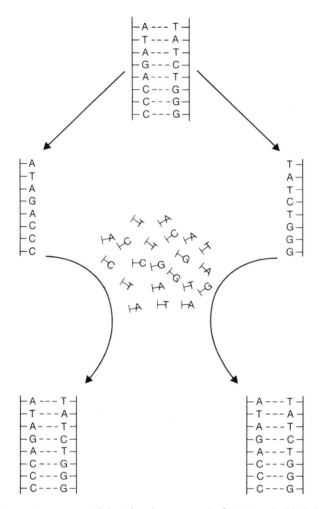

Figure 3.6   Anyone provided with only one strand of a DNA double-helix and a supply of nucleotides (bases linked to atoms that form the backbone of DNA) could easily re-create the original double-helix by following the rules of base-pairing.

triphosphates', consisting of one of the four bases linked to a sugar group and three phosphate groups. As a DNA molecule undergoes replication it is unwound by the activity of specific proteins. This exposes each single DNA strand to enzymes which catalyse the reaction in which appropriate nucleoside triphosphates supply the

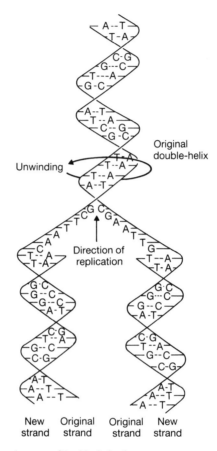

Figure 3.7   The replication of double-helical DNA.

linked base, sugar and phosphate groups (altogether known as 'nucleotides') needed to form a new complementary strand. Of course the enzymes are unthinking automatons which cannot consciously 'know' the rules of base-pairing, but their structure is such that they will only catalyse the joining reaction if the new base is able to pair up with its partner in order to form the double-helical structure. If a 'wrong' base binds to the catalytic site of the enzyme it will soon be jostled off the enzyme by thermal motion. Only when the 'right' base arrives will the enzyme catalyse the reaction which links it into the new DNA strand. So the combination of enzyme structure and the structure of the four bases ensures that the rules of base-pairing are automatically adhered to by the unthinking chemicals involved.

During DNA replication, existing single strands of DNA serve as 'molecular moulds' or 'templates' on which the required new strands can be formed. The shape of the base A in the 'mould' only permits the base T to fit into it as part of a new double-helix. The shape of T only permits A to fit, the shape of G only permits C to fit and the shape of C only permits G to fit.

So DNA molecules can become copied because of the chemical rules of base-pairing, which only allow specific base-pairs to be part of a double-helix; and because of the existence of enzymes which can link the appropriate bases (as parts of nucleotides) into new DNA strands.

How neatly our intuition sometimes fits with inner and hidden reality. Through the ages many proud parents have expressed pleasure at the desirable characters they feel their offspring share with them, by declaring such phrases as 'doesn't he really fit the mould'; and some men, on discovering that a child they thought was theirs was really the offspring of someone else, have bitterly declared that they 'always knew that he didn't fit the mould'. The analogy is so apt because deep down at the level of genes it is not really an analogy at all. If one could lift out the DNA from a child and snip it up into handy sized pieces, one would find that each single strand from each piece really would 'fit into the mould' of appropriate single strands of DNA taken from one or other of the parents. The DNA of a child could fit into the mould of the DNA of its parents, with all the As and Ts and Gs and Cs paired up in accordance with the rules of base-pairing, because a child's double-helical DNA really is formed on the mould of single strands of parental DNA. The only slight complication to be added to that image concerns the rare novelties called 'mutations', which cause the changes in the structure of DNA which allow species to evolve over many generations (discussed in chapter 9).

So much for the copying of DNA; what about its ability to direct the manufacture of specific proteins?

You should recall from chapter 1 that the first step towards protein maufacture is the formation of a single-stranded copy of the DNA of a gene in the form of RNA. This RNA copy is made in much the same way as new DNA (see figure 3.8). The DNA double-helix unwinds,

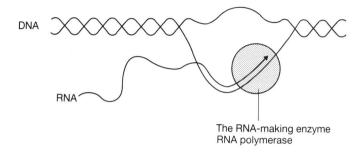

DNA

RNA

The RNA-making enzyme
RNA polymerase

Figure 3.8   DNA is transcribed into RNA when an enzyme uses one strand of the DNA double-helix as the mould or template on which a new and complementary strand of RNA is formed.

allowing enzymes to use one of the DNA strands as the mould or template on which a new and complementary strand of RNA is formed. RNA is virtually identical to DNA, so its formation is governed by the rules of base-pairing as before, although in RNA a base known as U (for uracil) takes the place of the very similar base T found in DNA. (The only other difference between RNA and DNA is that each nucleotide of RNA has an extra oxygen atom not found in DNA.)

The process in which an RNA copy of any gene is made is known as 'transcription', since the genetic message is being transcribed from the DNA original into an RNA copy.

Soon after it is made, the messenger RNA (mRNA) copy of a gene moves out into the cell cytoplasm, where it is used directly to govern the manufacture of a specific protein. To understand how it achieves this, you must remember that the difference between different genes, and therefore between different mRNAs, is simply that they have different *base sequences*; and the difference between different proteins is that they have different *amino acid sequences*. So an mRNA containing a specific sequence of bases strung out along its length must somehow direct the creation of a protein made up of a specific sequence of linked amino acids.

How is this conversion of specific structure brought about, all automatically and thanks only to the mindless workings of physics and chemistry? It occurs on the 'ribosomes' – molecular 'work-benches' composed of protein and RNA which are the chemical automatons which give nucleic acids the power to generate protein-based life.

As you work through the description of how a single specific protein molecule is made, remember that all the proteins involved in making a new protein must themselves have previously been constructed in exactly the same way, and all the RNA molecules must have been copied from the genes that encode them. The phenomenon of life is a self-sustaining global automaton in which proteins act on DNA to make RNA, and then on RNA to make new proteins, which can in turn take the place of the old proteins to act on DNA to make RNA, then more proteins, and so on *ad infinitum*. Also remember that, like everything else that occurs within the cell, all the steps of protein manufacture are essentially just chemical reactions and interactions taking place between all the proteins and RNAs and other molecules, ions and atoms involved. It would be difficult and tedious to describe all the chemical steps involved, many of them are not even fully known, but chemical reactions driven by the inevitable tendency of energy to disperse are believed to be the only phenomena involved in all the internal workings of life.

When a messenger RNA molecule meets a ribosome the two become bound together, and they then slide along relative to one another with the protein encoded by the mRNA being assembled as they go (see figure 3.9). The best way to understand how proteins are made on a ribosome is actually to forget about the ribosome itself first of all. The ribosome can be regarded as the work-bench on which all the necessary tools (mainly enzymes and other proteins and RNAs) and raw materials come together. This work-bench is formed when its component proteins and RNA molecules spontaneously aggregate together as a result of the chemical interactions between them, but it is what *happens* on this work-bench that really matters.

In the cell cytosol, surrounding the ribosome, there are lots of small RNA molecules known as 'transfer RNAs' (tRNAs). They are called transfer RNAs because these are the molecules which actually transfer individual amino acids into the structure of a growing protein chain. Each transfer RNA carries a specific amino acid which becomes linked to one end of the transfer RNA by the action of enzymes within the cell. The opposite end of each transfer RNA contains a group of three

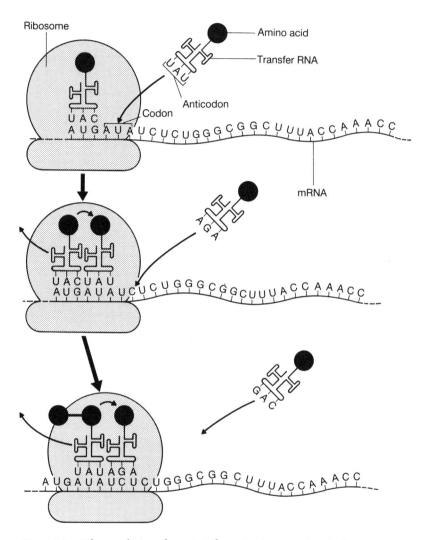

Figure 3.9   The translation of genetic information into proteins. A ribosome moves along mRNA, allowing transfer RNAs to bind to successive codons and let the amino acids carried by the tRNAs be linked together into a protein chain. Linkage of the first two amino acids of a protein is shown. Repetition of this procedure allows a long mRNA to direct the manufacture of a long protein chain.

exposed bases, known as an 'anticodon', which can form base-pairs with any three complementary bases, known as a 'codon' on a messenger RNA. So each tRNA carries a specific amino acid, and also contains an anticodon able to bind to a particular group of three bases on mRNA. That is how tRNAs manage to bring about the incorporation of amino acids into a protein chain under the direction of mRNA: each time a set of three bases on mRNA becomes exposed at a specific site on a ribosome, a matching tRNA binds to these exposed bases and allows the amino acid it carries to become part of a protein chain.

So a ribosome 'marches' along an mRNA, exposing a specific sequence of mRNA codons and allowing appropriate matching anticodons of tRNAs to bind to the codons and bring their amino acids as they do so. As the amino acids are brought to the ribosome in this way, enzymes link them up into a protein chain.

So by the time the ribosome has travelled the full length of the mRNA a protein chain will have been constructed whose amino acid sequence depends entirely on the base sequence of the mRNA. A sequence of bases in mRNA will have been 'translated' into a sequence of amino acids in a protein chain. The process is known as translation because it translates genetic information from the language of nucleic acids (DNA and RNA) into the language of proteins.

Once a protein chain has been completed it is released from the ribosome, and is then able to fold up into the precise conformation which allows it to perform some useful task, such as acting as an enzyme, a structural protein, a transport protein, or whatever. The DNA of a gene will have given rise to a messenger RNA, which will have given rise to a specific protein encoded by the gene. The whole process, involving both transcription of the gene into mRNA, and translation of the mRNA into protein, is known as gene 'expression'. By being transcribed into RNA and then translated into protein, genetic information can gain active expression in the form of working protein molecules.

That brings us full cycle in this brief summary of the central mechanism of life. We began with protein chains, each consisting of a specific sequence of linked amino acids. We saw how these protein chains spontaneously fold up into very specific three-dimensional conformations, and then considered how these folded proteins can perform various tasks required for the survival of a living cell. Some of the proteins in each cell catalyse the vital process of replicating the cell's genetic material – DNA – and copying that genetic material into messenger RNA molecules, one mRNA being generated for each gene. We then saw how these messenger RNAs travel out into the cytoplasm, become bound to ribosomes, and then bring about the manufacture of specific proteins when specific transfer RNAs, each carrying a specific amino acid, become bound by base-pairing to the successive sets of three bases, called codons, on mRNA.

In overseeing the manufacture of proteins, nucleic acid (this time in the form of RNA) is again acting as a molecular template or mould. It provides a three-dimensional surface into which only certain transfer RNA anticodons can fit (by forming base-pairs), and since each transfer RNA also carries a specific amino acid, the three-dimensional shape of the RNA mould determines the amino acid sequence (and therefore the structure and all the activities) of the protein it gives rise to. Notice that the fundamental chemical principle of selective binding lies at the heart of nucleic acid function as well as of protein function. The selective binding of the base A to the base T, and of the base G to the base C, underpins the structure of the double-helix and allows the double-helix to be automatically replicated. It also allows different codons on a molecule of mRNA to bind to specific anticodons of tRNA and hence direct the incorporation of specific amino acids into a growing protein chain.

Proteins do the work – genes generate the future

From what I have said so far it is clear that life is a result of the *mutually interdependent* interaction between two classes of chemical – proteins and nucleic acids. The importance of the proteins is that they catalyse the chemical reactions which together constitute life, and they also form much of the physical structure of life. The importance of the nucleic acids is that they determine what sorts of protein molecules are brought into being. Life would appear to consist of an endless cycle in which proteins make new genes and genes make new proteins, the new proteins continually take over from the old ones (since all individual molecules are mortal and eventually become damaged and degraded) and make new genes, which make still more proteins, which make yet more copies of the genes. ... And, as this cycle turns around and around, the proteins also construct and maintain all the parts of the delicate and fabulously complex creations we call living cells and organisms.

The story so far suggests that genes and proteins are both equally important, equally significant to the maintenance of life – and so they are; but why, then, are genes so often acknowledged as playing the dominant and most important role? Why do the schoolbooks all tell us that genes are the 'genetic blueprints' of life, the 'software' of the living world, the molecules which determine what sort of living things the world contains. There is a good reason for accrediting some funda-mental importance to our genes, and it will be emphasized in a moment, but it is not because our genes are any more vital than our proteins. Genes need proteins and proteins need genes – each is equally dependent on the other; but there is something very special about our genes none the less.

What is special about our genes (or about nucleic acids in general) is that they alone have the potential to branch out of one organism's gene–protein cycle and start up a new one, and they alone have the potential to *change* in a manner which allows the phenomenon of life to 'evolve' from one type of life into another. Let us look in more detail at what I mean.

Genetic material is special because it can become faithfully replicated, because it governs the specificity of the chemical reactions of life, and because it can allow changes to that specificity to be perpetuated.

By becoming replicated, genetic material gives rise to the copies of itself that are needed to set up new generations of life. The replication is catalysed by protein molecules, but it is the genetic material that is actually replicated and then passed on to new cells and new organisms. Genetic material is also the repository of the chemical specificity of life, because it is the molecular mould which determines the sequence and shape of all the proteins used by life. These two aspects of genetic material come together to give it the ability to permit life to evolve. Changes in the base sequence of DNA can be perpetuated through future generations, because the altered DNA can become replicated. Thus if the base sequence of a gene changes (we will see how in chapter 9), that change can be passed on to subsequent generations. The effect of the change will be felt at the level of proteins, because an altered gene gives rise to an altered protein, but it is the DNA of the gene which actually holds and preserves the change. So the only way the shape of life can change – the only way single cells can evolve into magnolias and monkeys and men – is by the slow accumulation of changes in genetic material.

If a protein molecule undergoes some spontaneous change, by reacting with some chemical for example, that might well alter the activity of the protein – it might even make the protein better at its job – but there is no way for the change to be passed on to future generations. The altered protein molecule will eventually be degraded, unlike an altered molecule of DNA which can be replicated to create many more copies of itself.

So the main reason why genes are often considered to be the 'master molecules' of life, despite the fact that the proteins do most of the chemical work, is because changes to genes not only change the actual DNA affected and the proteins it codes for, they can also be passed on during DNA replication to bring about permanent changes to life. Change a gene and you can change both the gene and the protein it codes for, for ever. Change a protein, and you are changing only the protein, and the change lasts only as long as the individual molecule which was changed.

The base sequence of genetic material such as DNA is often referred to as genetic 'information', because it determines what sort of proteins there will be, and therefore what sort of life there will be. To say that genes contain the genetic information of life is just another way of saying that they determine the chemical specificity of life. While proteins get on with the work of constructing and maintaining all organisms, the genes carry the chemical information that will generate the life of the future.

When one organism gives rise to another organism, it does so by passing a copy of all its genetic information into a cell, or group of cells, which can break away from the parent to become a self-sustaining living thing on its own. There is no need for the parent to pass on all its proteins to its offspring, because the genetic information can be used to generate all these proteins, provided the molecular machinery (ribosomes, transfer RNA, some enzymes etc.) is available to allow the genetic information actually to be used to create protein.

That is the story of life based on genes that encode proteins. The

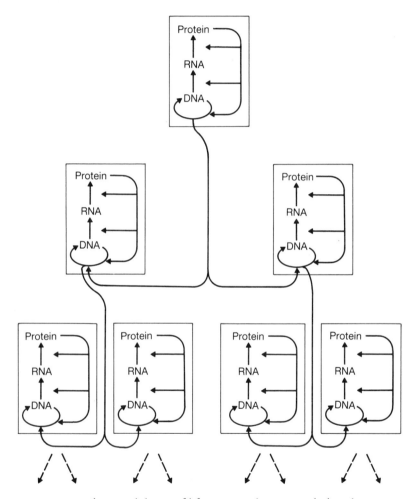

Figure 3.10   The central dogma of life: DNA makes RNA which makes protein, which makes DNA make RNA and protein. Nucleic acids (DNA and RNA) are as dependent on proteins as proteins are dependent on nucleic acids; but nucleic acids (especially DNA) can pass on the genetic information needed to generate new generations of life.

genes and the proteins are mutually interdependent, neither one is 'more important' than the other; but the proteins do the work while the genes generate the future (see figure 3.10).

Some complications

The best way to describe any complex system is first to of all expose the essential central principles governing its structure and operation, and then to explore any details and embellishments to whatever extent is necessary.

In this chapter we have looked at the essential principles of what has become known as 'the central dogma' of molecular biology. In its most terse form that dogma states that 'genes make RNA, which makes protein.' A fuller version might state that 'specific sections of the nucleic acid DNA, referred to as genes, give rise to RNA copies of the genes, thanks to the catalytic efforts of many proteins; the RNA copies then interact with the molecular machinery of the cell, particularly its ribosomes and transfer RNAs and proteins, to direct the production of specific new protein molecules; and the protein molecules of the cell, which are all made in this way, catalyse all the chemical reactions of life, including the ones involved in making new proteins.' More comprehensive versions would describe the way in which DNA molecules can become replicated to provide the copies of the genetic information needed by future generations; and then would explore the structure and chemical activities of DNA and RNA and proteins, until arriving at a version very similar to the contents of this chapter so far.

The version in this chapter so far, however, is still a very simple and general one which omits a great many details and complications. Since this is a book of essential principles, rather than of details, that need not concern us too much. You could stop reading this chapter now and still understand almost all the rest of this book. I hope you will not stop reading now, however, for it will be well worth while to pause for a moment, look back at the central dogma, and consider a few complications and further details. These are embellishments upon the simple central principle that genes make RNA, which makes protein, rather than completely new principles.

From what has been said so far you might have a picture of an organism's genome (i.e. all its DNA), as simply a long piece of DNA which is subdivided into many genes, with one gene following on directly after another (figure 3.11a). The first complication to disturb that simple picture has already been referred to in chapter 1. The genomes of most organisms, and all 'higher organisms' such as ourselves, consist of several separate segments of DNA, known as chromosomes. So most genomes do not reside on a single double-helix, but are split up into several chromosomes (figure 3.11b).

The second complication is that in the DNA of most organisms the genes do not follow on from one another as directly as figure 3.11b suggests. There are usually 'gaps' between the genes, with a gap simply being a region of DNA which is never copied into RNA and does not code for a protein (figure 3.11c). Some organisms contain a great deal of this 'non-coding' DNA, especially the so-called higher organisms such as ourselves. It has been estimated, for example, that less than 10 per cent of human DNA actually codes for functional proteins or RNAs, a fact which makes the human genome look a bit like a large notebook with only a few pages used, and with lots of empty spaces available for future exploitation. It is tempting to think of all the non-coding 'spacer' DNA as useless and redundant 'junk', and much of it probably is useless and non-functional, but things are certainly not as simple as that. Some of the non-coding DNA is undoubtedly involved in allowing the expression of nearby genes to be properly co-ordinated and controlled. For example, it includes sites of specific DNA sequence to which specific proteins can bind and influence the expression of nearby genes. It probably also includes sites which have a critical influence over the way in which the DNA becomes folded up into the tightly packed structure of the chromosome (see p. 81).

So, much of the DNA of many organisms does not encode protein and is not copied into RNA. Some of that DNA plays a vital role in controlling the expression of the genes and the folding of the DNA, but much of it probably is 'blank space' between the meaningful messages we call genes.

Further complications and surprises appear if we look more closely at the structure and activities of genes and the RNAs they give rise to. The RNA molecule which is initially manufactured by direct copying

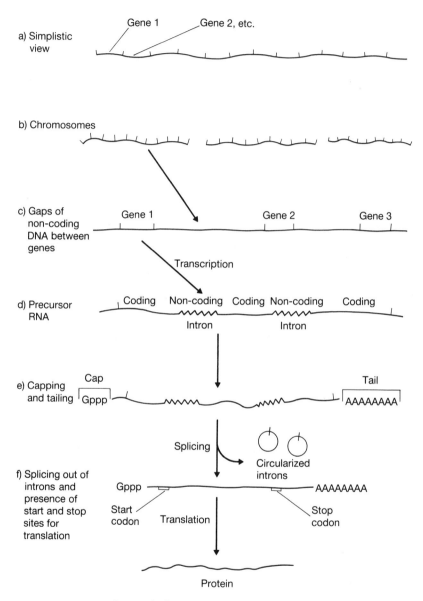

Figure 3.11   Some factors which complicate the structure and function of the genomes of many organisms (see text for details).

of a gene is not identical to the mRNA that ends up bound to the ribosome and being translated into a new protein. Instead, it is a 'precursor' RNA which must undergo various important chemical modifications. A rather simple chemical 'cap', for example, is added to the start of the RNA, while a long 'tail' consisting of many copies of the nucleotide *A* is added to its end (see figure 3.11d–e).

The most significant modifications to the precursor, however, involve the removal of specific portions from the interior of the RNA molecule, and the joining together of the remaining portions into mature mRNA (see figure 3.11e–f). This 'splicing' process occurs because the genes of many organisms include stretches of DNA which, although copied into the precursor RNA, do not code for any part of the protein made by the gene. So these 'intervening sections' or 'introns' must be spliced out of the RNA before it becomes the mature mRNA which is used to code for protein. In many cases the splicing process is catalysed by specific enzymes which neatly snip out the unwanted RNA and rejoin the wanted bits. Some introns, however, are able to fold up spontaneously into a structure which catalyses its own removal from the precursor RNA. Such 'self-splicing' introns provide a rare example of a catalytic process within the cell which is not mediated by protein, but by RNA.

A few other examples of RNA molecules which can act as simple catalysts have turned up. These do not threaten the central dominance of proteins as the molecules which 'do the work' within living things, especially since the catalytic powers of the proteins are required to make every type of RNA. Catalytic RNAs are an interesting exception to the rule that proteins are the catalysts of life. They are possibly remnants of an earlier age, right back at the origin of life itself, when the catalytic powers of the proteins may not have been 'discovered' and RNA may have served as a simple catalyst assisting the emergence of the chemistry of life.

While we are on the subject of RNA, it is worth while to point out that a few genes do not code for protein molecules at all, but simply encode various functional RNA molecules, such as the RNAs which form an integral part of the ribosome. In addition to these ribosomal RNAs, various other small RNA molecules are made by the cell and then go on to perform important, but still not fully characterized, functions. So a further complication or embellishment of the central dogma is that some genes simply make RNA, and this RNA performs important functions in its own right, rather than directing the manufacture of proteins. Once again I must emphasize, however, that proteins are required to catalyse the formation of all such RNAs, so these RNAs are as dependent on proteins as is the rest of the cell.

Let us return to the mRNA molecules, now fully 'processed' by the addition of caps and tails and the splicing out of their non-coding introns. As I have already described, these mRNAs travel out from the nucleus, into the cytosol, to become bound to ribosomes; but the entire length of the mRNA does not code for protein. Instead, there is a sequence of three bases, known as the 'start' codon which marks out the site at which protein-coding begins. In other words, the start codon marks out the site at which the first tRNA molecule becomes bound to the mRNA and brings with it the first amino acid of the protein. As the ribosome travels along the mRNA from this start codon, successive codons are exposed at the critical site on the ribosome surface, allowing appropriate tRNAs to bind to these codons and cause the appropriate sequence of amino acids to be linked up into the new protein chain. Eventually, some distance before the end of the mRNA, the ribosome will encounter a 'stop' codon. This is simply a specific sequence of three bases for which there is no corresponding tRNA. When the ribosome reaches this stop codon protein manufacture ceases and the new protein is released into the cytosol to fold up and get to work, as an enzyme, a structural protein, a transport protein, or whatever.

In many cases the new protein will not be able to perform its function immediately. It may need to combine together with other proteins into a functional multi-subunit complex. It may need to be chemically modified by the addition of various chemical groups such as phosphate groups or various sugars. It may need to associate with specific cofactors such as a particular type of metal ion or some more complex 'coenzyme' group. In all cases, however, the development of the new protein into a functional protein will either proceed spontaneously, thanks to the chemical activity of the folded protein and the chemicals it must interact with, or else it will be catalysed by various enzymes already present within the cell.

Finally, I want to return to the genome and its DNA to reveal one further complication or embellishment of the simple view of DNA which makes RNA, which makes protein. Everything that has been said so far has encouraged a view of DNA molecules as long wispy threads floating about in the nucleus. This is an utterly misleading picture. Real DNA in real cells is wound up tightly around an array of protein molecules, and then the complex of DNA and protein is wound up some more (see figure 3.12).

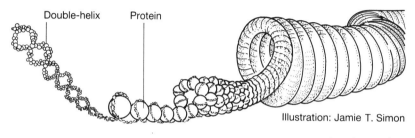

Double-helix   Protein

Illustration: Jamie T. Simon

Figure 3.12   The DNA of most cells is wound around proteins, and then the complex of DNA and protein is wound up further into the tightly packed structure of a chromosome. This is one idea of how the DNA and protein might be packed together.

One of the reasons for this dense 'packing' of DNA is simply the problem of getting all the DNA required by a cell into the small space of the nucleus. Another reason may be the need to keep careful control over which genes are active within any type of cell, and which are inactive. It seems likely that, for a region of the genome to become active (for its genetic information to be 'expressed' as RNA and protein, in other words), the densely packed structure of the chromosome must 'loosen up' or become 'unpacked' in some way. So a more accurate view of your genome is one in which long stretches of the tightly packed structure of figure 3.12 are disrupted at intervals by regions in which the DNA and proteins have become unpacked and therefore exposed to the enzymes of the cell which copy the DNA into RNA. I should emphasize, however, that much remains unknown about DNA packing and the way in which that packing is disrupted when genes become active.

So various complications and embellishments make the cell a more complex place than the central principles might suggest. We have looked at a few of the complications, but there are many more. They are all interesting and important, but they should not be allowed to obscure our view of the essential central simplicity of the chemistry of life: genes make RNA, which makes proteins, which make us all work. That is the conventional way of stating the central dogma. From the point of view of the proteins it can also be stated in the form of the title of this chapter, which tells us that proteins make genes make proteins. These two versions really say the same thing, of course, because proteins are as dependent on the genes that encode them as genes are dependent on the proteins that catalyse their replication and their copying into RNA.

There is a vital division of labour between the interdependent genes and proteins though: proteins are the molecular 'labourers' of life, while genes are the molecular 'manuals' which store the information needed to make new generations of protein labourers, and new generations of the cells and multicellular organisms which these labourers create. Proteins do the work while genes generate the future.

4 The powers of proteins

The story of the assembly of complex living things from simple chemicals available in the environment is essentially a story about the activities of the proteins. When told that living things are largely constructed and maintained by the spontaneous activities of proteins – a single class of chemical compound – many laypeople might react with confusion and disbelief. How can a single class of chemical compound 'do enough' to create a living cell? How can it do enough to create me? Any confusion and disbelief can only be encouraged by the realization that the proteins which create all these wonders are constructed out of only 20 simple chemical building blocks – the amino acids – each consisting of between 10 and 27 individual atoms. Complexity derived from simplicity indeed!

Proteins do lots of different things to construct life, but the great variety of things that proteins do can be divided into nine main categories:

Proteins can act as enzymes – molecules with a specific folded structure which allows them to bind to particular chemicals and encourage these chemicals to undergo specific chemical reactions.

Proteins can play structural roles – forming much of the basic physical framework of living things.

Proteins can form contractile assemblies – having the ability to undergo cycles of contraction and relaxation, giving cells and multicellular organisms the ability to move.

Proteins can act as transporters – selectively binding to certain chemicals in some environments, and releasing them in other environments.

Proteins can act as chemical messengers – being manufactured in one location, and then travelling to another site where they can bind to some other molecule to cause some appropriate chemical response.

Proteins can act as the 'receptors' of chemical messages – selectively binding to messenger molecules and responding to the arrival of the message by bringing about specific changes to the chemistry of the cell.

Proteins can act as chemical 'gates' and 'pumps' – controlling the passage of other chemicals through some barrier, either by simply opening and closing chemical 'channels' through a barrier, or by actively pumping chemicals from one side of a barrier to the other.

Proteins can act as chemical controllers – controlling the activities of other proteins, and of DNA and RNA, by becoming bound to the chemicals they control and altering their activities in some way.

Proteins can act as defensive weapons – binding to foreign organisms or diseased cells to initiate a series of events leading to the neutralization or destruction of the 'target'.

That is an impressive list, although still quite a short one to summarize the main activities of the chemicals that 'do the work' of making life.

Proteins can do all these things because their amino acid sequences cause them to fold up into structures which can selectively bind to other chemicals, and encourage things to happen to these chemicals once they have become bound. We have already looked at the factors involved in the necessary selective binding – the surface of a protein contains folds and cavities and clefts into which only the chemicals with which the protein must interact can fit. The forces which hold proteins and the chemicals they interact with together are usually weak forces of electromagnetic attraction between various parts of the proteins and other chemicals involved; but they can also involve the squeezing of hydrophobic chemicals into clefts away from the surrounding water, and the formation of covalent bonds.

What happens after the selective binding has taken place? In general terms two things can happen, sometimes both. First, the chemical interaction between the protein and the chemicals it binds to may encourage a specific chemical reaction to occur. The products of this chemical reaction will then be jostled away from the protein by thermal motion and an act of enzymic catalysis will have taken place. Secondly, the act of binding to other chemicals may cause a disruption in the structure of the protein as it experiences new electromagnetic pushes and pulls due to the arrival of the new chemicals on its surface. A slight movement of the protein chain near to the site of interaction can be transmitted through the entire protein until a 'conformational change' has taken place which leaves the protein with a new overall three-dimensional conformation. The chemical activity of the protein in this new conformation may be different from its activity in the original conformation. Thus proteins can be switched from one chemical state to another by the binding of other chemicals to their surface. In its original state, for example, a protein might have no enzymic activity, and so will be 'inactive'. The binding of some specific chemical may then subtly alter the protein's conformation so that a catalytic site appears on its surface and the protein becomes an active enzyme.

So there are three central principles underlying the many activities of proteins: they can selectively bind to specific chemicals; they can catalyse chemical reactions involving the chemicals which they selectively bind to; and the act of selective binding can induce conformational changes in the proteins which can alter their chemical activities in various significant ways (see figure 4.1).

Selective binding, catalysis and induced conformational change are the three unifying simplicities to look out for whenever you contemplate the powers of the proteins.

Enzymes make the chemistry that sustains life happen

Thousands of chemical reactions take place within living cells, each one catalysed by a particular enzyme. Enzymes cannot make reactions

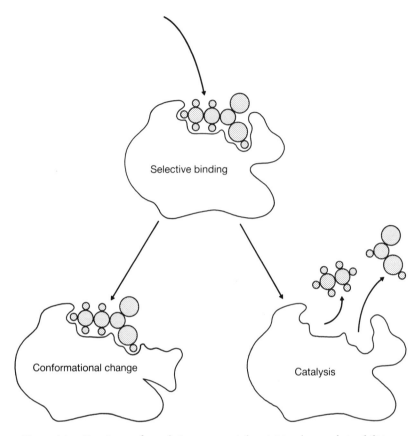

Figure 4.1   Proteins perform their many varied activities due to their abilities to participate in selective binding followed by catalysis and/or conformational change.

occur which could never occur without the help of the enzymes. In other words, enzymes cannot make otherwise impossible chemistry become possible. What they can do is dramatically *speed up* the rate at which specific chemical reactions take place (the term 'catalyse' effectively means 'speed up').

If all the chemicals present in a living cell were gathered together, *apart from the enzymes*, then an enormous number of different chemical reactions involving these chemicals would be theoretically possible. Many of these reactions really would proceed without the help of enzymes, although mostly at imperceptibly slow rates since the chemicals involved would need, by chance, to bump into one another with appropriate energy and in appropriate relative orientations, and so on. Many other theoretically 'possible' reactions would not proceed at all, because the activation energy needed to make them go would be too great. In other words, the random thermal motion of the molecular mixing bowl would be insufficient to cause the chemicals to collide with sufficient violence to make the reactions proceed. Also, the chances of all the required chemicals bumping into one another at the same time or in some appropriate sequence might be too small. Out of all the possible reactions between these chemicals of the cell, including both the ones that could proceed without enzymic help and those that could not, only a small fraction would be reactions that are actually used by and needed for life.

Returning all the enzymes of the cell to this dead chemical 'soup' would bring about an astonishing transformation. Suddenly the specific reactions which the enzymes could catalyse would begin to proceed at incredible rates – up to many thousands of times faster than they would have proceeded in the languid chemical soup (if they were able to proceed in that soup at all). The random thermal jostling of the molecular mixing bowl would still be required to allow the reacting chemicals to meet up with the appropriate enzymes, but the enzymes could bind to and hold on to each type of chemical until all the other required chemicals were met up with in turn. Once an enzyme had met and captured the required starting materials (also known as 'substrates'), the very special environment of the enzyme's surface would encourage the reaction between the chemicals to proceed at a great pace. If the enzyme catalysed a major and complex reaction, then further substrates might need to bind to the enzyme and react with the products of the first stage of the overall reaction, and so on through further steps; but eventually the reaction catalysed by the enzyme would be completed, leaving the products free to fall off the enzyme and leaving the enzyme free to meet up with another set of starting materials to perform the cycle of catalysis once again.

So enzymes have the ability to bring order and efficiency to a randomly reacting chemical soup. By selectively catalysing some reactions, while giving no help at all to many other possible reactions, they channel the chemical potential of their surroundings along very

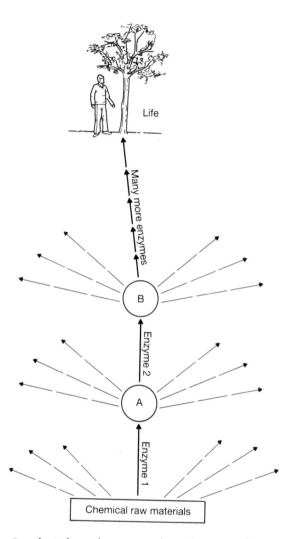

Figure 4.2   By selectively catalysing some chemical reactions (bold arrows), while giving no help at all to many other possible reactions (dashed arrows), enzymes channel the chemical potential of the environment along the narrow chemical paths that lead to life.

narrow and defined paths – the paths that lead to life rather than random chemical chaos, to growth and evolution rather than decay (see figure 4.2).

One vital point about the activity of enzymes needs to be stressed: they do not catalyse chemical reactions in any preferred direction. In other words, if an enzyme exists which can catalyse the reaction of A and B to form C and D, it will also be capable of catalysing the reverse reaction, in which C and D combine to form A and B. So under appropriate conditions in which A and B are abundant around the enzyme while no C and D are present, the enzyme will steadily convert A and B into C and D; but under other conditions, in which C and D are abundant while A and B are absent, the enzyme will catalyse the conversion of C and D into A and B. Enzymes are *facilitators* of chemical change, but they cannot influence the nature of that change. They can speed up the rate at which some chemical mixture reaches the point at which the forward and reverse reactions are proceeding at equal rates, and therefore nothing is changing *overall*, but they cannot influence the balance of chemicals which constitutes that final 'equilibrium' state. The final equilibrium state of a reversible chemical reaction depends on the nature of the reacting chemicals themselves.

For many reversible chemical reactions the equilibrium state, in which forward and reverse reactions proceed at the same rate, arises when there is a massive imbalance of the chemicals on one 'side' of the reaction over the chemicals on the other. It may occur, for example, when C and D are 10,000 times more abundant than A and B. In other cases, however, things may be more delicately balanced, with the equilibrium state occurring when the chemicals on both sides of the reaction are present in roughly equal abundance. In all cases an enzyme merely speeds up the movement of the reaction towards that equilibrium state. Enzymes speed up the possible, rather than making possible what was previously impossible.

In a book such as this you are not going to find a description of all the chemical reactions which take place within living things, and all the enzymes that catalyse these reactions. That would confuse the intended audience rather than enlighten and would cover many more pages than I have available (many of the reactions are not even fully known); but we can consider in a few sentences some of the main chemical tasks which are accomplished within living things.

Since living things grow and reproduce they must obviously gather up from the environment the chemicals needed as the raw materials for that growth and reproduction. They often need to convert these raw materials into more suitable forms before they are actually made use of; and then when they are made use of they must be correctly incorporated into the new substance of the organism – new genes and new proteins, new membranes and ribosomes and so on. At the same time, all the waste products of an organism's metabolism must be 'made safe' and got rid of. Many of the chemical reactions involved in sustaining life are ones which would not normally proceed within some dead chemical soup, simply because they involve converting raw materials from low energy states into high energy states; and so in a dead soup they would tend to go in reverse, if anything, with much of their energy being dispersed into the environment. So all living things need chemical mechanisms which allow them to drive low energy raw materials up into the high energy state associated with proteins, nucleic acids, and many of the other chemicals of life; and these chemical mechanisms must proceed spontaneously, without violating any of the laws of chemistry and thermodynamics. These mechanisms all involve cells somehow 'capturing' some of the energy available in their environment, and using it to power all the energy-requiring reactions within the cell. The subject of the energy transactions which power the growth and reproduction of life is so important it has been given its own chapter (chapter 6), but for the moment we should note that many of the enzymes of life are involved in the essential mechanisms of energy capture, and storage, and supply.

So the main tasks which must be accomplished by the enzymes of life concern the processing of raw materials available from the environment, the incorporation of these raw materials into the chemical fabric of life, and the capture, storage and use of the chemical energy which is required to drive many of the reactions of life forwards. The ultimate effect of all the chemistry of life is simply the growth and reproduction of the organisms concerned.

All that living things really do is grow and reproduce. They may develop large and complex bodies to assist them in this task, with eyes and ears and hair and teeth and brain and bone; and they may develop a variety of complex behaviours and even civilizations as they go; but in essence they are on this earth because they can grow and reproduce.

Life first arose and then evolved because any simple and spontaneously created chemicals, or networks of interacting chemicals, which were able to grow and reproduce would do just that – they would grow and reproduce and so multiply and evolve until they or their descendants covered the earth. Everything living things do is rooted in the requirements for growth and reproduction.

Enzymes are the chemicals which catalyse all the chemistry which makes growth and reproduction take place. We can find a good and very important example of how enzymes manage to bring living order to the chemistry of the earth by examining one of the most fundamental processes for all life – the incorporation of carbon atoms into the living substance of a cell.

All life on earth is based on carbon. Chains of bonded carbon atoms form the chemical 'skeletons' of most of the molecules of life. Look at the atomic structure of DNA, RNA, proteins, fats, carbohydrates, most vitamins and most of the myriad other chemicals found in any living thing, and you will find carbon atoms in great abundance, linked, of course, to lesser numbers of various other types of atom such as hydrogen, oxygen, nitrogen, sulphur and phosphorus.

So where do these carbon atoms come from? Ultimately, they come from the carbon dioxide gas in the earth's atmosphere and dissolved in its oceans, and the creatures responsible for trapping the required supplies of carbon are the plants (including tiny 'plankton') and 'photosynthetic' bacteria. Plants and photosynthetic bacteria capture the carbon atoms needed by all life. Other living things then get their supplies of carbon either by eating plants or photosynthetic bacteria, or by eating other animals which have grown by eating plants or photosynthetic bacteria.

When carbon dioxide gas permeates into a plant and then becomes dissolved in the fluid within the plant cells, it can meet up with and bind to an enzyme with the grand title of 'ribulose diphosphate carboxylase'. This is the enzyme which traps the carbon and makes it available as the ground substance of life. The enzyme catalyses the addition of carbon dioxide to a molecule known as 'ribulose

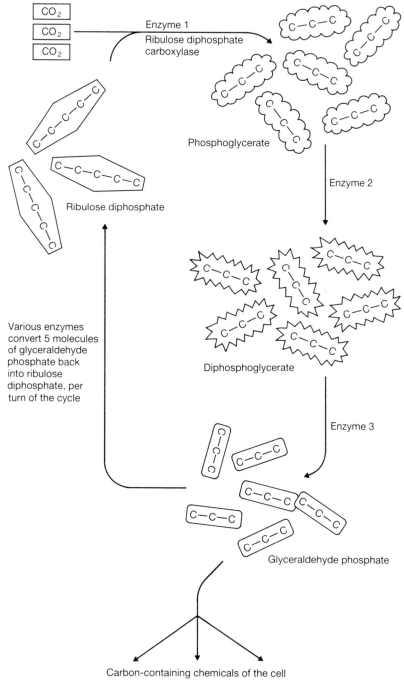

Figure 4.3   The carbon fixation cycle of plant cells.

diphosphate', and then splits up the product into two molecules of 'phosphoglycerate' (see figure 4.3). The names and the precise chemical structures are unimportant. What is important is that each time this enzyme goes into action a new carbon atom is converted into a chemical form which can be utilized by the cell. The starting materials or substrates of the reaction are carbon dioxide, which contains one carbon atom, and a ribulose diphosphate molecule containing five carbon atoms. By the time the reaction is over, all six carbon atoms are to be found distributed between two molecules containing three carbon atoms each.

It has been estimated that the enzyme ribulose diphosphate carboxylase may be the most abundant protein on the earth. It is certainly one of the most important, because almost all forms of life rely, directly or indirectly, on its carbon-trapping powers to gather up the carbon needed to let that life live and grow.

So what happens to the molecules of phosphoglycerate churned out by this enzyme, and where does the cell get its supply of ribulose diphosphate needed to keep the carbon-trapping reaction going? The answer can be seen in figure 4.3, which shows you how a plant cell continuously captures carbon in a self-sustaining 'metabolic cycle'. This cycle, known for obvious reasons as the 'carbon fixation cycle', operates like this: three molecules of ribulose diphosphate are each individually worked on by the ribulose diphosphate carboxylase enzyme to generate a total of six molecules of phosphoglycerate. These are then individually converted, by another enzyme, into a slightly different form; and then another enzyme converts them into molecules of 'glyceraldehyde phosphate'. The glyceraldehyde phosphate is a central metabolite of all life. It can be acted on by a variety of different enzymes which allow it to be used as a raw material for the construction of many types of chemicals needed by the cell, such as sugars and fats and amino acids. So by incorporating carbon atoms into glyceraldehyde phosphate, the enzymes of the carbon fixation cycle make these carbon atoms available for the manufacture of sugars and fats and amino acids, and ultimately of proteins and DNA and all other carbon-containing chemicals needed by the cell.

Out of every six molecules of glyceraldehyde phosphate produced during each turn of the carbon fixation cycle, only one is actually available for use as a basic raw material. The remaining five are used to regenerate the ribulose diphosphate needed to allow another turn of the cycle to proceed. Thus various enzymes convert five molecules of glyceraldehyde phosphate back into three molecules of ribulose diphosphate. These three molecules of ribulose diphosphate can then react with a further three molecules of carbon dioxide, ultimately to generate another molecule of glyceraldehyde phosphate for use as the basic carbon-containing raw material of all life.

Looking at the way in which life grabs hold of carbon atoms demonstrates many important points about the chemistry that allows life to work. First, the chemistry of life is not simple. Living things work thanks to many different chemical reactions which are combined in many rather complex ways; but, *in principle*, what happens overall is very simple indeed: a series of enzymes binds to appropriate substrates and converts them into appropriate products, and each product can become the substrate of the next enzyme in a series. At the end of each series of chemical reactions a raw material, such as carbon dioxide, will have become incorporated into some ultimate product needed by the cell. As chemicals are converted from one form into another they are said to pass along 'metabolic pathways', although cyclical pathways, such as the carbon fixation cycle, are also known as metabolic cycles.

All the various metabolic pathways and cycles of a cell are interconnected, through chemicals common to different pathways and cycles, to form one grand overall scheme which is called the 'metabolism' of the cell. The precise chemical details of the total metabolism of a living cell are extremely complex, and in many cases not fully known, but remember, the overall principles of what metabolism involves are very simple.

So we have looked in outline at what enzymes do that makes them so important – they catalyse the chemical reactions which make the chemistry of life work. Remember that as far as we know life simply *is* chemistry. In other words, life is simply the overall result of many thousands of chemical reactions proceeding at appropriate times, in appropriate places, at appropriate rates. Biochemists admit to no mysterious essences or principles, which must be added to our knowledge of chemistry to explain how a living cell manages to live – its life appears to be simply the end result of all the chemical reactions proceeding within it, each catalysed by an appropriate enzyme. It remains possible that there may be some mysterious principle behind the origin of consciousness and thought and our apparent free will, and that principle might either be physical or spiritual in origin; but, on the other hand, there might be no such hidden principle – most scientists do not seem to believe there is. If we restrict ourselves to understanding, or at least describing, what happens to make a living cell live, leaving aside for the moment the problems of consciousness and thought and free will, then the cell appears to be merely a complex chemical automaton which makes the enzymes which make it work and which allow it to grow and make further enzymes. . . .

Structural proteins hold life together

The basic unit of living things is the cell, but cells need to be held in shape and held together and much of this holding and support is performed by proteins that play structural roles. Bone, connective tissue, skin and hair and nails are all given strength and substance by the structural proteins they contain.

Many of the structural proteins are long stringy molecules, often wound together into strong cables or fibres. One of the most important is 'collagen' which has a triple-helical structure in which three distinct helical protein molecules are intertwined. In other structural proteins various sections of a protein chain are held together to form flat sheets. The interior of the cell is criss-crossed by a network of structural proteins which is known as the cytoskeleton. The long protein 'bones' of this skeleton are formed by the spontaneous aggregation of many individual globular protein molecules. There is actually a wide variety of structural proteins found in many different locations, but, whether they are fibres, sheets or dense collections of globular proteins, they all serve to hold things together and give shape and substance and hardness to the structures of life.

Of course, what actually does the holding is the electromagnetic force. Structural proteins are held together, and are held in contact with other structural proteins or other structural components of the cell (such as cell membranes or complex structural carbohydrates) by the attractive power of the electromagnetic force. Cells use many strong chemical 'pillars' and 'beams' and 'glues' and 'cements', both inside them, to hold the internal structure of cells together, and outside of them, to hold different cells together; but the electromagnetic force is the fundamental 'glue' upon which they all depend.

Contractile proteins let life move

Structural proteins are often referred to as the molecular scaffolding of life, and the analogy is quite apt since so many structural proteins are long fibres or rods; but we think of scaffolding as a static, unchanging framework. Imagine, however, a structure built of scaffolding in which some of the scaffolding rods were able to slide past one another and then hold the whole framework in new positions. We would then have a dynamic scaffold able to change shape and to move. Some rod-like proteins are able to form just such a dynamic scaffold within the cell. They are sometimes called 'contractile' proteins, since they allow cells and muscles to contract; but the contraction is largely achieved by rods of protein *sliding* past one another, rather than by any individual proteins undergoing contraction.

Take a look at figure 4.4 to see what I mean. It shows how proteins allow your muscles to work. The two most important proteins of muscle action are 'actin' and 'myosin'. An individual actin molecule is a globular protein, but many such molecules can spontaneously assemble into long actin filaments. This assembly process can be recreated in the test-tube. Myosin is a bit more complex, it consists of six separate protein chains intertwined and aggregated together to produce a long rod-like 'tail' and a twin 'head'. Many such individual myosin units can then combine together to produce a myosin filament.

In muscle, actin and myosin filaments are enmeshed together as shown in figure 4.4. When the muscle contracts the actin filaments are pulled past the myosin filaments. How this happens is outlined in the figure. The myosin heads make contact with and bind to the actin filament. They then undergo a conformational change which causes the attached actin filament to be jacked inwards. The myosin head is then released until it reattaches elsewhere on the actin filament and

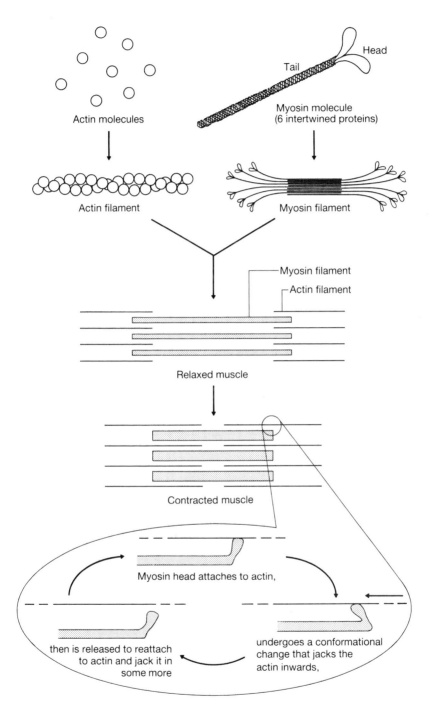

Figure 4.4   *The molecular basis of muscle contraction.*

jacks it back some more. This is happening right now within the cells of your eye muscles as your eyes scan back and forth across the page.

We will refer again briefly to muscle contraction later, when it is time to consider some of the energy problems faced by living things; but for the moment we can see that proteins such as actin and myosin are able to act together to provide the contractive force needed to allow cells and organisms to change shape and move from place to place; and remember that actin and myosin are mere molecules able to achieve such feats by virtue of their chemical structures, which allow them to aggregate into filaments and muscle fibres and to undergo the necessary changes in conformation when appropriate.

Transport proteins move things around

Many of the problems of organization are problems of transport, regardless of whether we are considering the organization of a cell, a human being, a school, a city, an army, a state, or even the whole of civilization. One of the main factors distinguishing complex organized systems from disorganized, chaotic ones is that things not only have their place, they also move about between various places in an organized manner. Modern society would collapse without the transport of raw materials and products from where they are available to where they are needed. The human body would soon die and disintegrate if its systems of transport collapsed. The most obvious system of bulk transport in the human body is the blood, which flows through our arteries, capillaries and veins like a 'river of life', bringing chemical raw materials (oxygen, water and food) to every cell of the body, and taking waste products away. Within this bulk system, however, the actual job of transporting specific substances is sometimes performed by small 'freighters' such as individual blood cells and even individual protein molecules. This is not to imply that all the materials transported through our bloodstream must be actively transported by cells or proteins, many are simply washed along in the stream, but some others do need some help, and in certain vital cases that help is given by proteins.

Oxygen, for example, is one of the most vital chemical requirements of the human body. In general terms we need oxygen, water and food in order to survive (we will consider why later). Most of us could survive without food for many weeks; we could last without water for a few days; but without oxygen we would die within a few minutes. Our bodies' supplies of oxygen are to be found in the cells lining the lungs, where the oxygen diffuses into the tissue from the atmosphere. The oxygen is needed by every cell in the body. It diffuses from the lungs into the bloodstream, but if it were simply carried along passively within blood, then the supplies of oxygen that eventually reached our cells would be grossly deficient. The oxygen needs some assistance in order to be efficiently transferred from the lungs to the rest of the body, and that assistance is given by a protein known as 'hemoglobin'.

The hemoglobin is found in our red blood cells. It is a multi-subunit protein composed of four distinct protein chains. Attached to each chain there is a cofactor called 'heme' which actually acts as the chemical vessel on which the oxygen is carried. At the centre of each heme group there is an iron ion to which oxygen can bind *reversibly*. In other words, when the concentration of oxygen around the heme group is high, an oxygen molecule will bind to the iron; but when the concentration of oxygen in the surrounding environment is low, then any oxygen bound to the iron will be released. This is the chemical basis of the ability of hemoglobin to act as an oxygen-transporting automaton. When red blood cells are passing through the tiny blood vessels lining the lung their hemoglobin molecules are exposed to high concentrations of oxygen, and so they become loaded with oxygen molecules attached to the iron ions of the heme groups. The flow of the bloodstream then takes the oxygen-rich red blood cells away from the lungs and around the body. When they enter tissues in which the oxygen concentration is low, and which therefore need further supplies of oxygen, the required oxygen is automatically off-loaded from the hemoglobin. The deoxygenated hemoglobin is then swept around the bloodstream within its red blood cells until, once again, it arrives in the oxygenated tissues of the lungs where it picks up more oxygen and the cycle of oxygen transport begins once more.

Actually, hemoglobin is a rather more subtle and efficient oxygen transporter that the above outline would suggest (see figure 4.5). The multi-subunit structure of hemoglobin allows it to load up with

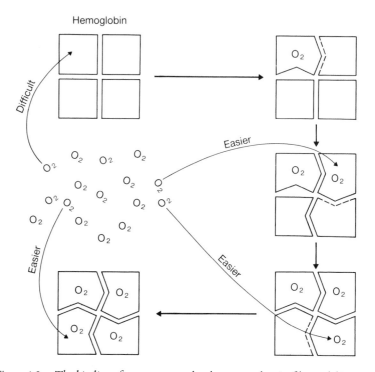

Figure 4.5   The binding of one oxygen molecule to one subunit of hemoglobin causes conformational changes within the multi-subunit complex which makes the binding of further oxygen molecules easier. The diagram is highly schematic - the detailed mechanism of the process is still unclear.

oxygen extremely efficiently in the oxygen-rich tissues of the lung, and offload it extremely efficiently in tissues requiring oxygen. This is because, when one of the subunits picks up an oxygen molecule, it causes the conformation of the hemoglobin complex to change slightly, in a way which allows further oxygen molecules to bind to the other subunits much more easily. So the binding of one oxygen molecule to one subunit of an empty hemoglobin complex greatly encourages the binding of oxygen to the other three available sites. This makes the multi-subunit hemoglobin complex a bit like a four-seater car in which the first person into the car unlocks the doors for another three passengers. The crucial step in loading the car is getting the first person in, after which the first person helps all the others to climb aboard.

An opposite effect occurs when loaded hemoglobin reaches a tissue in need of oxygen: the loss of one oxygen molecule from one subunit causes a conformational change in the complex which allows the other three oxygen molecules to be off-loaded much more readily. A suitable analogy to this would be an unstable four-man boat, since, if one man jumps overboard, he may rock the boat sufficiently to make the other three fall out!

So the structure of the subunits of hemoglobin allows the multi-subunit complex as a whole to undergo conformational changes which ensure that, in oxygen-rich tissue, the binding of one oxygen molecule to one subunit encourages the binding of oxygen molecules to the other subunits; and, in oxygen-deficient tissue, the departure of one oxygen molecule from one subunit encourages the departure of oxygen from all the other subunits. This allows hemoglobin to load up with oxygen very efficiently in the lungs, where oxygen is plentiful and is therefore likely to meet up with the hemoglobin and become bound to it, and then unload very efficiently in the tissues that need oxygen, in which the scarcity of oxygen encourages it to be lost from the hemoglobin, rather than taken up by it. It is a nice example of the subtle chemical behaviour that is possible in a world in which proteins do most of the work.

Messenger proteins and receptor proteins let one cell influence another

Messages are as vital to organized systems as transport. You only need to imagine the chaos that would befall a country if its telecommunications and postal systems ground to a halt to realize the importance of messages; and imagine if people also became unable to communicate directly, by word of mouth, by notes, by sign language, by expressions or whatever. Human society would collapse without communication, just as each individual human life would soon end if the body's cells were unable to communicate with one another.

What do we mean when we refer to cells 'communicating'? In

essence, we mean that the chemical activity of one cell is able to alter the chemical activity of other cells in some consistent and meaningful fashion. Cells achieve such chemical communication in various ways, but the most vital way is by releasing chemical 'messenger' molecules (the biological equivalent of the postal system, if you like analogies), and many of these messengers are either proteins, or small fragments of proteins.

The principles of intercellular communication using messenger molecules are very simple (see figure 4.6). The molecules are produced in one cell, released from that cell, and then travel (via simple diffusion or carried within the bloodstream) to other cells. The cells which must receive the message – the 'target' cells – will have special proteins, often protruding from their surface, which are able to bind selectively to the messengers they must receive. These proteins of the target cells are called 'receptor' proteins. The binding of messenger to receptor then initiates some chemical disturbance within the target cell, and this disturbance is the response to the message which the messenger carried. The binding of the messenger, for example, might alter the conformation of the receptor, causing it to catalyse some change within the target cell which might cause it to start growing. So the message carried by the messenger molecule in this case would have effectively been 'start growing', and the binding of the messenger to the receptor would have ensured that the message was not only

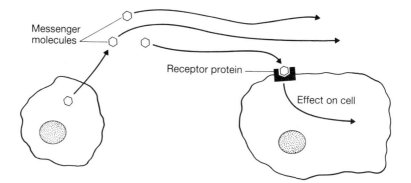

Figure 4.6   Messenger molecules (often proteins or peptides) released from one cell can bind to specific receptor proteins of other cells and bring about some specific effect on these cells.

received, but was acted upon. A biological messenger molecule is more like a legal summons than a friendly note or some junk mail advertisement – it commands the target cell to react in a precise way to the arrival of the message.

One familiar example of a protein messenger is the hormone 'insulin'. This is manufactured and secreted by cells in the pancreas, and it then travels around the bloodstream and binds to a wide variety of cells to cause them to initiate a variety of different responses. (One messenger molecule can elicit different responses from different target cells, just as one message, such as 'there's going to be an election' would elicit different responses from politicians, journalists, stockbrokers, bookmakers and the general public.) The best known effect that insulin has on cells is to encourage them to take up supplies of glucose from the bloodstream. A deficiency of insulin, found in sufferers of diabetes, causes the levels of blood glucose to rise dangerously unless the insulin deficiency is made good by injections of artificial supplies.

There are many other protein messengers, which cause cells to respond in a wide variety of different ways. Lots of messenger molecules are composed of only a few amino acids linked together, rather than the hundreds of amino acids of a typical protein, and such small proteins are distinguished with the separate name of 'peptides'. Some important messenger molecules are merely single amino acids. Many other messengers are completely different chemicals from proteins, peptides and amino acids, so proteins certainly do not carry all the messages that pass between cells, but message-carrying is yet another important process which proteins and peptides can perform. All other message carriers, of course, are created by the catalytic effects of enzymes, so proteins underpin the abilities of all messenger molecules, even when they are not the messengers themselves; and all the receptors for messenger molecules are proteins (usually glyco-proteins).

Proteins can act as gatekeepers of the cell

Import and export are further activities vital to the economy of the cell. Chemical raw materials must enter cells, and waste products must get out. The selective entry and exit of chemicals into and out of the cell also underpin the electrical activity of the nervous system which we presume to be responsible for consciousness and thought (see chapter 8).

All cells, of course, are surrounded by the barrier of a fatty membrane. Some chemicals are able to pass freely through this membrane, so they automatically diffuse from the side in which they are in high concentration to the side where they are more scarce. Many more chemicals, however, are completely unable to cross the cell membrane, or any other biological membrane, unless they are given some help. The required help is most often given by specific types of proteins which are found embedded in the membrane (see figure 4.7).

Some membrane proteins form tiny pores or channels which permit particular chemicals, of appropriate size and charge, to pass through the membrane by simple diffusion. Other proteins play a much more active and specific role in membrane transport, by binding to specific chemicals on one side of the membrane and transporting them to the other, where they are released. This transport across the membrane is probably achieved in several different ways, depending on the proteins concerned. In some cases the binding of the substance to be transported will cause a conformational change in the protein, causing the substance to be transferred to the other side. In a few cases the proteins may somehow revolve through the membrane to release their cargo on the other side. Regardless of the details, proteins can act as very selective and efficient 'gates' allowing specific chemicals through the membrane, while giving no help at all to other chemicals.

In some cases these membrane proteins merely allow chemicals to be transported down a 'concentration gradient'. In other words they allow chemicals to move from the side of the membrane in which they are present in high concentration, to the side where they are more scarce. This is a situation very similar to simple diffusion, although it is called 'facilitated diffusion' since it needs the help of protein to facilitate it. In both simple diffusion and facilitated diffusion chemicals can never be transported from a region of low concentration into a region of high concentration. In other words, transport mechanisms based on diffusion can never serve to *concentrate* any chemical within a cell, or to expel any chemical from a cell if that chemical is more abundant outside than inside. Diffusion simply allows the chemicals

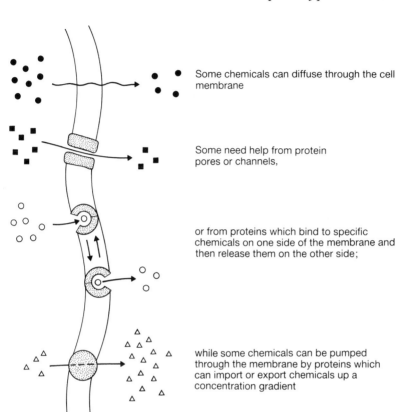

Figure 4.7   Ways for chemicals to move through the cell membrane.

concerned to flow across the membrane, from regions of high concentration to regions of low concentration, until the concentrations become equal on both sides. Once equal concentrations of any chemical have been attained on both sides of the membrane diffusion stops.

Cells can, however, actively import or export many chemicals against a concentration gradient. In other words they can accumulate high concentrations of chemicals even though they are very scarce outside the cell; and they can efficiently expel chemicals into an environment in which the chemicals are already far more concentrated than they are inside the cell. The proteins which perform these feats are not gates, but 'pumps', but in general terms they act in much the same way as before, binding to a chemical on one side of the membrane, and then undergoing some conformational change which results in the chemical being transferred to the other side. The difference is that the operation of such pumps requires the expenditure of chemical energy. In other words, for the pumps to work their operation must be coupled to chemical reactions which drive the pumping mechanism forward. We shall consider how such 'chemical coupling' of energy-requiring reactions to energy-releasing ones can be achieved in chapter 6, when the whole topic of the energy changes associated with life is considered.

For the moment, we have met another vital role which proteins can play in the mechanisms of life. They can lie embedded in biological membranes and control the passage of chemicals from side to side – either by forming pores or channels, or by binding to specific chemicals and directly mediating their transfer to the other side.

Regulatory proteins can keep control

A living cell is a self-regulating molecular automaton. The various parts of the cell are dependent on one another, and the complex network of chemical interactions between these parts keeps the whole cell living and growing as it should. Obviously, then, the various parts and components of the cell must be able to influence one another, and some of the major mediators of this influence are regulatory proteins. Regulatory proteins are simply proteins able to bind to other components of the cell and modify their behaviour (see figure 4.8).

Many proteins regulate the behaviour of other proteins. A protein-regulating protein is one that can bind to a specific protein and alter it in some way which brings about a specific change in the target protein's function. The regulatory protein may, for example, catalyse some chemical modification of the target protein which causes it to change from an inactive into an active form, or from an active into an inactive form, or from a weakly active into a very active form, or from a very active into a weakly active form, or from a form in which it catalyses one reaction into a form in which it catalyses a slightly

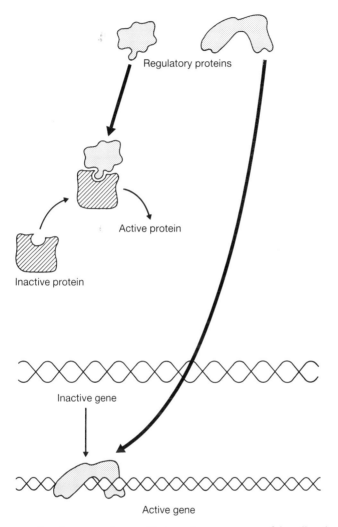

Figure 4.8   Regulatory proteins can bind to other components of the cell and modify their behaviour.

different reaction. There are endless possibilities, but the essential point is that some proteins can influence the activity of other proteins in many different ways. In many cases a protein which has been jolted into activity by the attentions of one regulatory protein, can then act as a regulatory protein itself and jolt a further protein into activity; this third protein may then activate another, and so on down a long biochemical 'cascade'.

Regulatory proteins can also influence the activity of the other great class of chemicals at the heart of life – the nucleic acids DNA and RNA. For example, many gene-regulating proteins exist which are able to bind to specific regions of DNA and either activate or inhibit the activity of the genes nearby. Some proteins, in other words, can bind to a gene, or close to a gene, and cause it to begin to be copied into mRNA (by a separate RNA-producing enzyme), while other proteins can achieve the reverse, and completely shut down the activity of a gene. The precise way in which gene-regulating proteins work is often unclear, and probably very varied. Many gene-regulating proteins probably bring about conformational changes in DNA, which convert it into an active or inactive form by making it either easier or more difficult for the proteins and other molecules needed for gene activity to bind to the DNA and 'get to work' on it. The details do not matter for this present brief review of the powers of the proteins. Proteins have the power to regulate the activity of other proteins and of nucleic acids, and that gives them the power to act as vital controllers of the life of the cell.

Proteins can be defensive weapons

Everybody has heard of antibodies, the chemicals which can protect us from specific infectious diseases. Antibodies are simply specialized protein molecules. They are manufactured within white blood cells and released into the circulatory system of blood and lymph, as well as into various other body fluids. An antibody molecule has two main parts of prime importance. First, it has a highly specific binding site which is able to bind to proteins and other chemicals (known as 'antigens') found on the surface of invading micro-organisms. Each different micro-organism carries different antigens on its surface, and so the body must manufacture different antibodies to attack them. Secondly, each antibody carries a 'tail' region at the opposite end to the binding site, which assists in the removal and destruction of whatever the antibody has become bound to. It works like this: if some micro-organism, say a virus, invades the body, it should eventually become covered with antibody molecules able to bind very specifically to the antigens found on the virus's surface. Once coated with antibodies, the virus may be effectively neutralized and unable to bind to and enter any of the body's cells. Even if it is not neutralized directly by antibody binding, however, the outlook for any virus covered in antibody should be bleak. The tails of the antibody molecules are able to bind to proteins carried on the surface of special scavenging cells known as 'phagocytes'. When they bind to the phagocytes, of course, the antibodies carry their bound virus with them, and the phagocytes are then able to engulf and digest both the virus and its surrounding antibodies.

So antibodies are defensive proteins which can bind to invading micro-organisms and either neutralize them directly or encourage their destruction by the body's defensive cells. There are other, related, defensive proteins which help the body to fight off infection and disease in other, rather more complex, ways. Once again, we have met a further great power of the proteins – by selectively binding to foreign invaders they can act as a very effective means of defence against disease.

This has certainly not been an exhaustive summary of the powers of the proteins, but it has introduced you to many of their most important abilities. The diversity of things which the proteins are able to do is bewildering, but remember that in general they achieve all their effects in rather simple ways. First, they selectively bind to the specific chemicals which they can act upon, or which can act upon them. Secondly, they either catalyse some chemical alteration of whatever they have become bound to, or they may undergo some conformational change which alters their activity in some significant way. Selective binding, catalysis and induced conformational change are the three central processes which underpin all the diverse powers of the proteins; and all three are brought about by the 'mere chemistry' which causes the atoms and molecules of the proteins and their substrates to interact and react in the ways in which the laws of physics compel them to do.

5   Membranes and metabolites

Living things can be composed of one cell, several cells, or a great many cells. There are many different types of cell in the living world. Figure 1.1 showed us the main components common to all animal cells, including many creatures, such as single-celled protozoa, which you might not normally consider to be 'animals'. Plant cells differ from animal cells in two important ways – they have a tough 'cell wall' composed of complex carbohydrates such as cellulose and found outside the cell membrane; and they contain bodies called 'chloroplasts' which, as we shall see in chapter 6, allow them to trap the energy of sunlight in the process known as photosynthesis. The other main type of cell is the bacterial cell. The cells of bacteria are much smaller and simpler than most animal or plant cells, and they lack many of the components present in these larger and more complex cells. Bacterial cells have no nucleus and no internal membrane-bound bodies like the mitochondrion, the lysosome, etc., of figure 1.1, but they operate according to the same central principles as other cells: they contain DNA which makes RNA which makes the proteins which make the cells work.

So, if we can explain how cells such as the one shown in figure 1.1 manage to live, and then also explain how such cells can become specialized and can co-operate together within a multicellular organism, we will have a good overall view of how chemistry makes life live.

We have already examined the most vital central mechanism of the living cell. This mechanism is essentially the interdependent activity of DNA and RNA and proteins: DNA molecules encode specific protein molecules, via RNA intermediaries, and these protein molecules then catalyse all the chemical reactions required to construct and maintain the cell, including the reactions which maintain DNA, replicate it when required, copy it into RNA when required, and allow the RNA to direct the manufacture of proteins. I have now told you several times, and in several ways, that DNA makes RNA, which makes protein, which makes cells, including their DNA and RNA and protein: that is the central principle of life.

But you are unlikely to be satisfied with an explanation as simple and succinct as that. Looking at figure 1.1 probably makes you want to know more about many of the bodies shown – the cell membrane, the mitochondrion, and so on; and you probably want to know how the working of the central mechanism allows all the required chemistry of the cell to occur in the right places, at the right times and to appropriate extents. If you tried, you could probably think of a host of problems which the workings of a cell must overcome in order to produce coherent life and growth – problems of raw material supply, of control, co-ordination and so on.

We will be looking at some of the details of the living cell and examining its problems over the next few chapters. In this chapter we will cast our gaze outwards from the central mechanism, in search of the simplicities which underlie the complex structure and operation of the rest of the cell.

In addition to its DNA, its RNA and its proteins, a cell contains two other main classes of components. First, it contains membranes, one of which surrounds the entire cell, while the others cordon off various parts of the cell into specialized compartments known as 'organelles'. Secondly, the cell contains a bewildering array of chemicals, which we can refer to collectively as 'metabolites', dissolved in the water of the cytosol or aggregated into large insoluble complexes. The metabolites are a rich mixture of chemical raw materials and stores; chemical intermediates formed during the manufacture of DNA, RNA, proteins and membranes; and a variety of small molecules which play important roles in allowing cells to work. So, *in principle*, there is really not much more to a cell than its central mechanism of DNA, RNA and protein, since, in addition to these 'big three', there are mainly just membranes and metabolites.

Membranes keep things in, and out, and apart

The basic structure of the membranes of the cell is blissfully simple – an inspiring illustration of the abilities of mere chemicals to act as the architectural automatons of life. The main components of cell membranes are chemicals known as lipids, a term which basically means 'fatty' substances that are insoluble in water. The lipids found in membranes all have the same general structure, consisting of a long fatty tail which is insoluble in water, and a short head group which is soluble in water (see figure 5.1). The tails are insoluble because they consist of chains of carbon atoms with hydrogen atoms attached

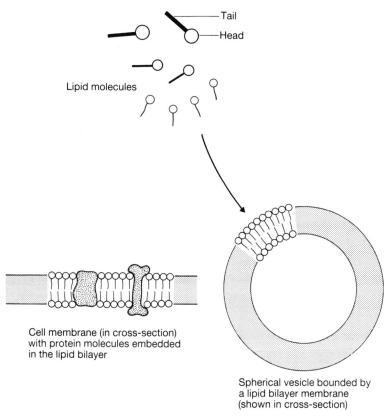

Figure 5.1   *Lipids and lipid bilayer membranes.*

('hydrocarbon' chains), with no regions of electrical charge which could interact with the slight negative charge found on the oxygen atoms of water, and the slight positive charge found on the hydrogen atoms of water. The head groups, on the other hand, usually do contain electrically charged regions which can interact with the charges on water molecules to form a stable and soluble structure.

These lipids, therefore, have split chemical personalities – the heads being strongly attracted to the watery environment both within and around cells, while the tails have no attraction whatsoever for the water, and so tend to be pushed away from watery regions wherever possible. How can the competing demands of this split chemical personality both be met at the same time? The answer is quite simple (see figure 5.1). These lipid molecules, if they are simply added to a beaker of water and shaken, for example, can spontaneously become organized into thin membranes in which the tail groups of the lipids are clustered together, away from the surrounding water, in the interior of the membrane, while the head groups are exposed to the water on both sides of the membrane. This happens automatically when the lipid molecules are pushed and pulled by the competing electromagnetic forces between the lipid molecules and water molecules, with the tail groups being squeezed out of the watery regions until the whole system settles down into the stable and low energy structure of the membrane.

Membranes like this are known as 'lipid bilayers', for obvious reasons, and the lipid bilayer is the basic component of all the membranes of the cell. It looks like a very organized structure, with all the lipid molecules neatly packed together in the correct orientation; but it is really a very simple structure which can form spontaneously, a bit like the organized structure of a crystal, whenever appropriate lipids come together in a watery environment.

In addition to providing the cell with an outer membrane which separates the cell from the outside world, lipid bilayers also form the boundaries of the various 'vesicles' (organelles) inside the cell, such as the nucleus, the mitochondrion, the endoplasmic reticulum, the Golgi bodies, the endosomes, the lysosomes and the chloroplasts of plants. These organelles are all specialized regions of the cell containing particular metabolites and proteins which co-operate together to perform particular specialized chemical tasks. The basic lipid bilayer of all the membranes of the cell is impermeable to most of the molecules found inside and outside of cells, so it forms an effective barrier which can keep some chemicals inside cells and some outside, and can separate some of the chemicals within a cell from all the others, by retaining them within some specialized organelle.

The organelles known as mitochondria house many of the chemical reactions which allow cells to convert the chemical energy of raw materials or stored chemical supplies into a more versatile and readily usable form. They will be discussed in more detail in chapter 6.

The nucleus, of course, is the organelle which contains the genome of the cell, along with a host of enzymes and other proteins which serve to maintain and package the genome and allow its genetic information to be used for the manufacture of proteins and RNAs.

The endoplasmic reticulum is a complex network of interconnected tubular channels and flattened vesicles which permeates the cell cytosol. Lots of proteins are manufactured by ribosomes which are attached to the membrane of the endoplasmic reticulum, allowing the proteins to pass quickly through the membrane into the interior of the endoplasmic reticulum, and so be separated from the rest of the cell cytosol. This is believed to allow specific proteins to be dispatched into various other organelles of the cell, or out of the cell. So the endoplasmic reticulum is a kind of molecular 'sorting office', with the proteins inside it or outside of it being dispatched to different locations.

The Golgi bodies are vesicles which break off from the endoplasmic reticulum, allowing them to carry the protein molecules they contain to various cellular locations, such as various other organelles, or to the environment outside of the cell.

The endosomes and lysosomes are vesicles packed with enzymes which are able to degrade or modify the other molecules of the cell. They form a kind of intracellular digestive system and 'breaker's yard'.

The chloroplasts are the specialized organelles of plant cells which house the chemistry of photosynthesis – the process by which the chemicals of living cells can capture the radiant energy of the sun and convert it into a usable form. As I have said already, we will be discussing what goes on inside the chloroplasts in chapter 6.

There are other types of organelles, although those listed above are the main ones. There would be little point in listing them all or in examining in great detail what goes on inside all of them. That would be the task of a cell biology textbook, not a simple introduction to life's most basic principles. The basic principle about all organelles is that they are small vesicles within the cell, bounded by lipid bilayer membranes and containing specific metabolites and proteins, and sometimes DNAs and RNAs as well. In each organelle a specific set of chemical reactions proceeds which helps the cell to process raw materials, create the substance of the cell, grow and reproduce, or capture the energy needed to power all these activities.

The various membranes which surround the cell and all its organelles are not composed simply of the basic lipid bilayer and nothing else. Instead, they all have various proteins embedded within them (see figure 5.1), which can give them further specialized abilities. Membrane proteins often float around within a sea of lipid (although they can also be 'held at anchor' in the one spot if required). Many of them are the selective channels, gates and pumps, discussed earlier, which allow certain molecules to pass through the membrane barrier, or to be actively pumped through. Other membrane proteins can give different membranes different structural properties, while others, which partially protrude from the cell surface, govern the interactions between neighbouring cells and allow cells to be held together. Different membranes also contain different types of lipids, causing them to have different properties of flexibility and permeability and so on.

Membrane biology is a very active and complex area of modern biological science, but, once again, the overall principles are simple: all biological membranes are based on the lipid bilayer – a stable sandwich of two layers of lipid molecules which can form spontaneously. The bilayer is impermeable to most molecules, though not all, but can be made selectively permeable to specific molecules by the presence of proteins embedded within the membrane. Other proteins can influence the structure and precise chemical behaviour of different membranes, and different lipids can create membranes with subtly different properties.

Metabolites – the molecular maelstrom

The interior of any cell is a seething 'metabolite pool' – a maelstrom of molecules rushing about and bumping into one another, meeting up on the surface of enzymes and reacting with one another, being released from the enzymes to bounce about some more until they meet up with the next enzyme or other type of molecule with which they must interact.

The chaos of a large department store on Christmas Eve, or during the January sales, is a reasonable analogy. There is order and logic within a scene of frantic and often seemingly chaotic activity. People bump into one another and push past one another and rush about from counter to counter with increasing panic in their eyes; but most of them eventually find what they want, and by the end of the day a fairly successful distribution of goods from the shop to the people who want them is achieved. Similarly, there is order within the seeming chaos of the molecular maelstrom inside the cell, and overall the successful operation of the chemistry of life is achieved.

All the diverse chemicals in the metabolite pool are ultimately derived from chemical raw materials which have been taken into the cell from the environment outside. As we have already seen, these raw materials can be small and simple molecules which enter the cell through pores and channels in the cell membrane, or which are passively or actively transported through the membrane by specialized membrane proteins. A few raw materials can get in by diffusing through the membrane directly, without any help from channels, pores or proteins. The mechanisms of raw material uptake considered so far, however, can only maintain a stream of small molecules into the cell, one at a time. Another much more dramatic mechanism can engulf large quantities of the exterior environment, or even large bodies as big as entire micro-organisms, and deliver them into the cell. This mechanism of large-scale raw material uptake is known as 'endocytosis'.

Endocytosis begins when a portion of the cell membrane buds inwards ('invaginates') and in so doing captures a great gulp of the exterior environment, perhaps including some large object such as a virus or a bacterium. As the 'endocytotic vesicle' buds further inwards, the membrane eventually completely surrounds the gulp of environment, and when this happens it can break off into the interior of the cell as a separate small vesicle. This occurs because, when two lipid bilayer membranes touch (or two different regions of the same membrane touch), they can automatically fuse together. So when the neck of the endocytotic vesicle is closed off the membrane can fuse and

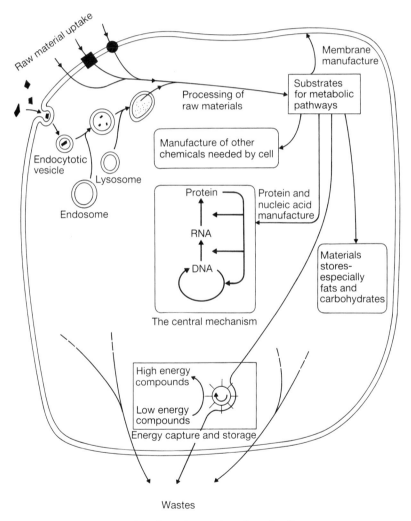

Figure 5.2   Highlights of the flow of metabolites through cells.

rearrange itself to yield a separated vesicle while leaving the cell membrane intact behind it (see figure 5.2). Once inside the cell, the endocytotic vesicle usually fuses with a cell body (another specialized vesicle) known as an endosome, and then with another called a lysosome. Inside the lysosome there are powerful enzymes which begin the task of breaking down the raw materials into simpler chemicals which can be used as starting materials for the manufacture of the things the cell needs to make.

Simpler chemicals which enter the cell directly may also be 'processed' in some way, although there is usually less need. Overall what happens to the raw materials taken into a cell is that they are processed by enzymes which convert them into the starting materials required to build new cells, unless they happen to be in a suitable form already. This degradation and processing of raw materials is part of a subdivision of metabolism known as 'catabolism' – the breaking down of chemical raw materials.

Another major aspect of catabolism is the continual breakdown of the components of the cell itself. Cells are not stable and unchanging structures like office blocks. Instead, most parts of a cell are in a state of continual demolition and renewal, known as 'metabolic turnover'. Imagine an office block in which a large team of builders is constantly moving through, knocking down existing walls and using the bricks to build up new ones; ripping apart the furniture and then reassembling it into new forms; peeling off the wallpaper, then using it as the raw material to produce new paper which is then put back up again; and all the time some new materials are arriving through the door, to assist in the continual rebuilding, while some of the older materials are constantly being discarded out of the windows. The living cell is in a very similar situation, with teams of enzymes constantly ripping down the structure of the cell while other teams of enzymes build it up.

This constant turnover might seem a very wasteful process, but it is absolutely vital to the survival of cellular life. It allows cells to adapt to changing circumstances by changing their shape and structure and overall activities very quickly, since the enzymes needed to break down the old and build up the new are on hand and active all the time. Life in the office block imagined earlier might sometimes be a little difficult and chaotic, but at least when change was required it could be brought about quickly, since the necessary tradesmen and supplies would always be on hand; and any mistakes made during the building process could always quickly be put right. Metabolic turnover bestows similar advantages on the living cell.

So the cell gets its chemical raw material supplies from the environment around it, and from the continual breakdown of existing cellular structures, and overall this destructive breaker's yard division of metabolism is known as catabolism. The part of metabolism which builds up new structures from the raw materials is called 'anabolism', and the complete metabolism of a cell is the result of its catabolism and anabolism combined.

The main components which the process of anabolism must produce are the cell's nucleic acids (DNA and RNA), its proteins and glycoproteins and other structural chemicals, its membranes and its stores of materials as insurance against any times when there is a scarcity of raw materials in the environment.

We have already seen how new nucleic acids are made, during either the replication of DNA or the manufacture of RNA copies of portions of DNA. In all cases a pre-existing strand of nucleic acid serves as a molecular template or mould which allows a new complementary strand of nucleic acid to be assembled on the molecular mould by the required enzymes.

We have also seen how new protein molecules are manufactured out in the cytosol, when the mRNA encoding a particular protein binds to a ribosome, which then travels the length of the mRNA and allows enzymes to join up the required amino acids (which are carried on tRNA molecules) into the correct protein. Many proteins are modified into glycoproteins when enzymes of the cell attach specific sugar groups to specific sites on the protein surface; and the presence of these sugar groups can critically alter the proteins' behaviour to allow the finished glycoproteins to perform specific and special chemical tasks.

Enzymes in the cytosol are also responsible for using simple raw materials to manufacture all the lipids found in the cell membrane. Once formed, these lipids become incorporated into appropriate membranes to sustain the turnover of the membranes of the cell and to allow the cell to grow bigger, and perhaps eventually to split into two daughter cells.

The major non-protein and non-lipid structural components of cells are various complex carbohydrates which, either on their own or attached to proteins or lipids, form many of the walls, cables, fibres, fillers, glues and pastes of life. These too must be manufactured at the right times and in the right places by the activities of appropriate enzymes.

When 'times are good' and cells have a more than adequate supply of raw materials available, any excess can be used to generate stores of some components, especially fats and carbohydrates. Enzymes can process many simple organic raw materials into fatty deposits which can be stored in the cell until required. Simple carbohydrates, such as molecules of glucose and other sugars, can be linked together into long chains of storage carbohydrates such as 'glycogen' (found in animal cells) or 'starch' and 'cellulose' (in plant cells).

So, at the chemical level, the interior of all living cells is a swirling sea of chemical activity. Thousands of enzymes are busy acting upon thousands or even millions of molecules of metabolites which are all passing along various complex and interacting metabolic pathways.

The breakdown and processing of raw materials, and the manufacture of nucleic acids and proteins and membranes and carbohydrates and storage molecules, forms just a part of the seething sea of metabolism. I have emphasized, so far, the many essential tasks that proteins perform within a cell, but smaller, simpler non-protein molecules also perform a great many vital tasks. These non-protein 'effector' molecules are all created by the enzymic activity of proteins, so my initial emphasis on the importance of the proteins is justified. That emphasis is also justified by the fact that most 'non-protein effectors' have their effects on protein molecules, or sometimes on nucleic acids. So they are created by proteins to bring about specific effects on the activity of other proteins or nucleic acids. They are vital intermediaries allowing proteins and nucleic acids to maintain their influence over the cell, but the proteins and nucleic acids are the chemicals which really matter. None the less, the range of different tasks performed by non-protein metabolites is bewildering. Some act as 'coenzymes', which become bound to enzymes and help them to perform their catalytic tasks. Others act as hormones, which are released from one cell to influence the activity of other cells; or as 'neurotransmitters', which are released from nerve cells to induce other nerve cells to transmit a nerve impulse (see chapter 8). Others act as localized 'tissue factors', released to control the activity of other cells nearby. A wide range of small non-protein metabolites act as internal signalling agents, binding to proteins, for example, and altering their activities in ways that allow all the various chemical activities in the cell to be co-ordinated in an appropriate manner.

Some vital non-protein metabolites act as a sort of 'energy currency' which serves to power all the cellular activities which might, at first sight, seem energetically unfavourable. What exactly do I mean by 'energy currency', and what are the energy problems faced by living things? This is a subject of great importance which has been largely ignored up till now, while an overall impression of the activities of life has been presented. The energy problem will now be dealt with all on its own, in a new chapter.

6 Getting the energy

We live in a universe which, at heart, may be very simple. It is a universe of space-time, matter, charge, force and energy. Everything we see around us, from the stars to the smallest visible living things, and everything outside the range of our vision, seems to be due to the interaction of space-time, matter, charge, force and energy.

One result of this fundamental simplicity is that the processes at the heart of the most complex forms of life are analogous to simpler but basically similar processes we can see every day in the world around us.

The world around us receives its energy almost entirely from the sun. If we watch the steam rising from a pavement after a rain shower, we see one of the effects of that energy – it heats up the molecules of water on the pavement, raising their 'energy level' until they have sufficient kinetic energy of motion to escape from the liquid water into the atmosphere in the form of water vapour. The sun pumps energy into the material of the world, raising that material to higher energy levels. What happens to the water molecules next? They rise as vapour, not only from wet pavements, but from streams and rivers and ponds and lakes and the sea, until, high up in the sky where the temperatures are lower, they lose their energy in collisions with slower-moving, colder molecules and atoms, causing the water molecules to condense back into tiny droplets of liquid water. We see the droplets as clouds, and very often we see them coalesce to form the raindrops which shower down on to the surface of the earth, to run off the sodden hilltops into streams and rivers and ponds and lakes and the sea.

So what is the overall effect of the flow of energy from the sun on the water of the world? It powers an endless cycle of evaporation, condensation, evaporation and condensation. . . . It raises cold water to the higher energy level of water vapour, and then, once the water vapour has lost its energy to the colder environment all around, some more of the sun's energy raises it up again to the energy level of water vapour.

Overall, energy is flooding from the sun and becoming absorbed by and distributed around the colder world of planet earth. Planet earth, in turn, is continually loosing some of it out into the colder world of outer space. The energy from the sun flows relentlessly into planet earth in accord with the irreversible dictum of the second law of thermodynamics, which tells us that energy must automatically disperse towards an even distribution. By now, billions of years after the earth was formed, the automatic and inevitable effect of the dispersal of the sun's energy has been to raise the energy level of the earth, and to keep it raised.

In considering the effect of the sun on the water of the earth, we can identify an obvious universal truth about the automatic dispersal of energy: in dispersing from one place to another, it raises the energy levels of the place where it is received. That might seem ridiculously obvious, but some people apparently have difficulty in accepting its implications for living things. If you examine the chemicals within living things you find that in many of them the constituent atoms are arranged in ways that are certainly not the lowest energy configurations possible on the earth. If all our nucleic acids, proteins, carbohydrates and lipids, for example, were to react with the oxygen of the air (effectively, to 'burn'), then all the atoms involved in the reaction would end up in lower energy configurations and a great amount of heat energy would be given out. So, many of the chemicals of life seem to be in rather high energy states, considerably higher than much of the environment around them.

This high energy status of life causes some people quite a problem. How, they ask, can unthinking physics and chemistry maintain the situation in which life is at a higher energy level than the environment all around it? How can a fertilized human egg cell possibly 'gather up' the energy and the matter from a lower energy environment to develop automatically into a highly organized living, thinking human? Surely the creation of a high energy organized human from the lower energy and disorganized materials of the earth is a violation of the second law of thermodynamics, which states that energy should disperse towards an even distribution, not become concentrated in small regions such as the human body?

People troubled by such questions should consider the waters of the world, the clouds and the rain. They should think of the billions of molecules of water which can automatically become gathered up from different bodies of water all many miles apart and then be deposited together in one small pool high up on a mountainside. The waters of the world are continually being raised up to the higher energy level of water vapour, and then deposited together in pools which each might seem to have been gathered up from a lower energy and more chaotic state into high energy organization. This happens, not in violation of the laws of thermodynamics, but as a direct result of the action of these laws. The same argument applies to living things: all living things are able to exist in their high energy organized state because the energy of the sun puts them there. People who think that life in some way violates thermodynamic law, are forgetting that the environment around living things is not only the 'cold' earth, it is the earth plus the sun and the continual supply of energy which flows from the sun. (Strictly speaking, of course, the environment around any living thing is the rest of the universe.)

When high energy and organized life is formed from lower energy and disorganized raw materials, it forms not in violation of thermo-dynamic law, but because thermodynamic law determines that the energy of the sun *must* disperse out to the earth and raise the energy level of the things that are found there. The raw materials of life are some of the things that are found there, and the energy from the sun raises these raw materials up into the higher energy levels associated with organized life, just as it raises water up into the sky and deposits some of it in tidy little mountain pools.

Water-wheels and reservoirs

To introduce you to the energy changes associated with the chemistry of life I have drawn an analogy between the creation of living things containing many high energy chemicals (i.e. those in which the electromagnetic force is resisted much more than it could be), and the raising of water vapour from the seas into the sky. We can continue with this analogy as we look deeper into the energetics of the living cell.

The raising of water to the skies is not an isolated and irreversible event, but part of a cycle in which the water eventually loses the energy gained from the sun and returns to the earth as rain, only to absorb some more energy and be lifted up once more, and so on. . . . Similarly, of course, the creation of a living being such as yourself is not an isolated and irreversible event, but is part of a cycle of life and death, of growth and decay. You are created, you arise as the 'dust' of the earth absorbs energy which has come, ultimately, from the sun, and then you die and decay back to dust and energy. In both the planetary water cycle and the planetary life cycle energy is taken in, to raise the water to the skies or the raw materials of the earth into life, and then given back to the environment as the water coalesces into rain or the life dies and decays (see figure 6.1).

One complication is that in any living thing the release of energy

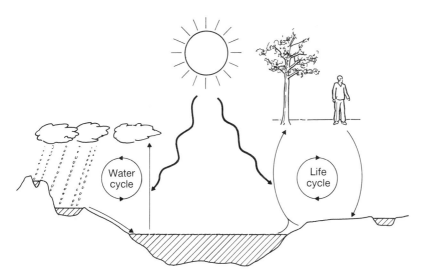

Figure 6.1   Both the planetary water cycle and the planetary life cycle are powered by energy from the sun.

back into the environment is proceeding all the time, as individual cells die and are replaced, and also as high energy chemicals either created by that life, or gathered up by it (as food), are allowed to participate in reactions which let them fall down into lower energy states.

The endless turning of the water cycle offers intelligent beings such as humans the opportunity to harness some of the energy flow which drives the cycle forward and put it to a variety of uses. As water rushes down a mountain stream, for example, it loses a little of the energy which it absorbed from the sun when it evaporated to the skies. Most of its energy was lost when it condensed back into rain, but when the rain fell on the mountainside a little of the energy was still trapped as gravitational energy, which is lost as the water rushes back to sea level. A man could build a water-wheel and dip it into the mountain stream, and the wheel would begin to turn. The power of the turning water-wheel could then be harnessed to do some useful work, such as milling grain.

A man who built his livelihood around the power derived from such a simple water-wheel might soon run into some problems. There would be times when he would have to sleep and, since he would be unable to attend to his mill, the water would continue rushing past him unused and wasted. There might be times when the stream ran dry, leaving him without any power to do his work, or times when the stream ran in rapid flood, providing much more power than he actually needed. There might be times when he had a lot of work to do, and could have used the power of several streams had they been available; and there might be times when work was scarce, causing water again to rush past unused and wasted.

He could solve many of these problems in a rather simple way, by constructing a pump which could be powered by his water-wheel, and which would raise water from the stream into a large storage tank, or even up into a reservoir in the mountainside. Then, when he was asleep, or the stream was in flood, or work was scarce, he could harness the power of the stream to build up a reservoir of water. The energy of this reservoir could then be utilized in several ways. First, of course, it could simply be used to power the original wheel when the flow in the stream was slack. Secondly, however, it could also be used to power many other wheels built on dry land, vastly increasing our man's overall work capacity, flexibility and the variety of tasks he could undertake (see figure 6.2).

If we look inside the chemical mechanisms of the living cell we find that they can harness the energy available in the environment, most of which ultimately comes from the sun, in a manner similar to the man who has built a water-wheel, a pump, a reservoir and many secondary wheels used to power many different tasks. Of course the energy reservoir of the cell does not consist of high bodies of water, but takes the form of high energy chemicals whose manufacture can be driven

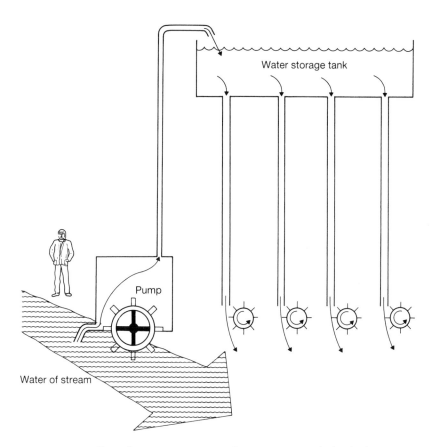

Figure 6.2   Cells can harness energy in a similar way to a man who has built a water-wheel to power a pump, which creates an energy reservoir, which can be used to drive many secondary water-wheels which power many different tasks (see text and figure 6.3 for details).

forward by the energy available in the environment; and whose degradation back into lower energy chemicals can then be used to drive forward all the energy-requiring reactions of the cell. In living things the roles of the water-wheels and pumps are played by various systems of proteins and membranes, while the most common immediate energy reservoir is a chemical known as 'adenosine triphosphate' (ATP).

The structure of ATP is shown in figure 6.3. It is most commonly manufactured from 'adenosine diphosphate' (ADP) and inorganic phosphate ions, whose structures are also shown in the figure. ATP is the cell's equivalent of water stored in a high level reservoir or a tank, because it takes an energy input to make it, while energy is given out when it breaks apart into ADP and phosphate.

Why should ATP have a higher energy level, or energy content, than ADP and a free phosphate ion? Remember the definition of energy given earlier as some sort of antiforce or ability to resist the power of a fundamental force. This tells us that systems in which there is a large amount of violation of or resistance to some fundamental force will be high energy ones, while those in which the force is less resisted or more satisfied will be lower energy ones. So, when ADP and a phosphate ion are combined together into the form of ATP, the overall resistance against the electromagnetic force must be greater than when they are in their free dissociated forms. Look, for example, at the line of three slight positive (δ^+) charges arrayed next to one another in the ATP molecule. The electromagnetic force causes these charges to repel one another, so the structure of ATP must involve considerable resistance to that force, a resistance needed to allow the molecule to hold together despite the 'efforts' of the electromagnetic force to break it apart. The ATP structure, in other words, is a high energy structure, in which the constituent atoms are arranged in a manner which involves considerable resistance against the electro-magnetic force. If the ATP splits up into ADP and a phosphate ion, one of the trio of δ^+ charges is removed, and products are formed whose structure involves less force resistance, or, in other words, less energy. This is a simplistic and only partial explanation of a rather complex situation, but it conveys the general principles involved.

The considerable resistance to the electromagnetic force embodied in the structure of ATP imposes a strain on the ATP molecule. It is like

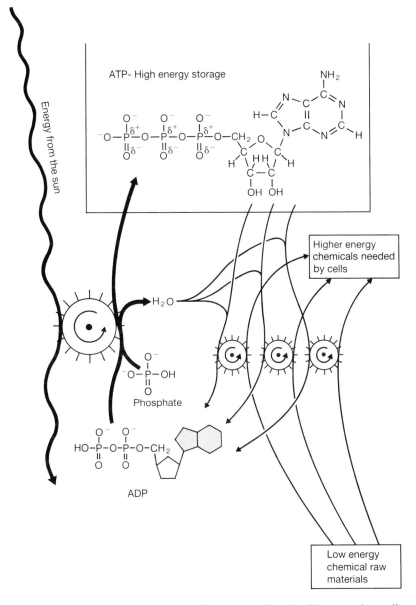

Figure 6.3   Cells can capture energy from the sun (directly if they are plant cells, indirectly if they are animal cells) and use it to make ATP. The energy stored in ATP can be used, when required, to power a wide variety of energy-requiring reactions in the cell.

the compressed spring of a jack-in-the-box just waiting to be released; and when it is released in some appropriate chemical reaction, then the energy level of the molecule falls as it splits up into ADP and phosphate.

Just as the force of water falling from a high gravitational energy level to a lower one can be harnessed to make various energy-requiring processes proceed, so the force of an ATP molecule falling from a high chemical energy level to a lower one can be harnessed to make a wide variety of energy-requiring chemical reactions proceed, as we shall examine shortly.

Given that ATP serves as an energy reservoir for the cell, how is it made? How do cells trap energy and use it to make ATP? The ultimate source of the energy which must be trapped or 'harvested' to generate ATP is usually the electromagnetic radiation from the sun. The creatures which perform the energy capture on behalf of most other creatures are the plants and some bacteria, and the energy trapping process is known as photosynthesis.

In plants, the chemistry of photosynthesis occurs in specialized organelles known as chloroplasts. As you can see from figure 6.4, a chloroplast is a complex multi-membraned structure enclosing three separate fluid-filled compartments – the inner 'thylakoid spaces', the 'stroma' and the 'intermembrane space'. The place to look for the chemicals that capture the energy of the sun is the inner 'thylakoid' membrane – a highly specialized membrane containing particular lipids which make it impermeable to ions. This membrane also carries various proteins and other chemicals which actually perform the chemical tricks of photosynthesis. First, there is the green pigment known as 'chlorophyll', a complex non-protein organic chemical. What is special about chlorophyll is that it contains electrons which can absorb electromagnetic radiation from the sun and, as a result, be 'kicked up' or 'excited' into a high energy state in which they are capable of being transferred on to other nearby chemicals, should such 'electron-accepting' chemicals be available. Embedded in the inner membrane of the chloroplast there is a whole series of molecules which can accept excited electrons from chlorophyll, and then pass them on to other members of what is known as an 'electron transport chain'. Many of the components of this chain are proteins.

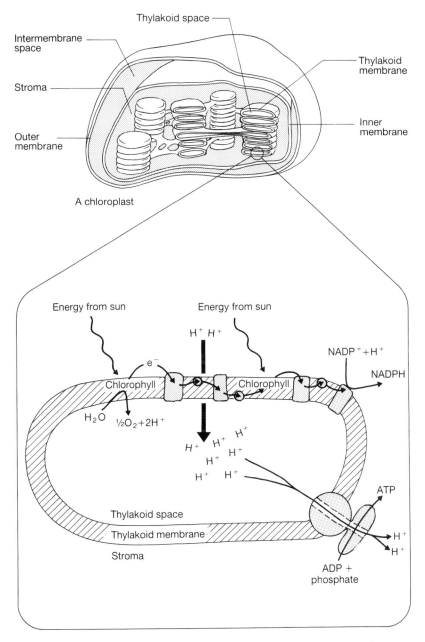

Figure 6.4   How chloroplasts capture the energy of sunlight and use it to make ATP and NADPH (see text for explanation).

So the energy of sunlight falling on the green leaf of a plant is trapped when it kicks an electron out of a chlorophyll molecule, and the electron is then channelled away by being passed from one member to another of an electron transport chain. This leaves the original chlorophyll molecule positively charged, but it can be regenerated by a reaction with water, during which electrons from water molecules are donated to chlorophyll molecules, leaving behind hydrogen ions and oxygen gas. This is the source of the oxygen gas generated by all photosynthetic life, and which we must breathe in order to keep ourselves alive (for reasons which will become evident later). Chlorophyll molecules actually appear at two sites in the electron transport chain, so there are two sites at which electrons can receive 'energy kicks' from the sun (see figure 6.4).

The ultimate fate of the electrons which are ejected from chlorophyll molecules is to combine with hydrogen ions and a chemical known by the abbreviation $NADP^+$. The reaction between these three species generates NADPH. We will see what happens to the NADPH shortly, but what about ATP, whose manufacture I set out to explain?

During the ejection of electrons from chlorophyll, and their transfer along the electron transport chain, hydrogen ions are released into the thylakoid space, and removed from the stroma, so, effectively, hydrogen ions are 'pumped' across the thylakoid membrane. Some hydrogen ions probably are literally transferred across the membrane, by the action of the components of the electron transport chain, while some are generated in the thylakoid space by the splitting of water, and used up in the stroma by the creation of NADPH from $NADP^+$ (see figure 6.4).

The precise details of the movement of the hydrogen ions are complex and the subject of some debate, but three essential facts must be grasped: first, during the passage of electrons along the electron transport chain an imbalance of hydrogen ions (a hydrogen ion 'gradient') is created across the thylakoid membrane; secondly, this gradient is a high energy situation, since it results in one side of the membrane becoming positively charged due to an accumulation of positive hydrogen ions, while the other side becomes, relatively, negatively charged (separating positive and negative charge despite the efforts of the electromagnetic force to pull them together obviously requires energy); and thirdly, this high energy hydrogen ion gradient is built up automatically as an inevitable consequence of the transfer of electrons along the electron transport chain, which is powered by the energy of the sun.

The structure of various components of the electron transport chain causes them, effectively, to transfer hydrogen ions across the membrane as they pass electrons down the chain – one process cannot happen without the other, and since the electron transport *must* happen, because it is being powered by the energy of the sun, so the hydrogen ion gradient must form in accompaniment.

So, what the complex system of membranes and proteins and other chemicals embedded in the thylakoid membrane achieves, is to store some of the energy coming from the sun in the chemical form of an ion imbalance across the thylakoid membrane. This initial and temporary energy store then serves to generate ATP as will now be explained.

There is an enzyme embedded in the thylakoid membrane which can manufacture ATP from ADP and phosphate ions available in the stroma; but it can only catalyse this reaction if it is accompanied by the effective transfer of hydrogen ions *through the enzyme* from the thylakoid space towards the stroma. In fact, it is the flow of hydrogen ions through the enzyme, powered by the hydrogen ion imbalance generated by the electron transport chain, which actually provides the energy needed to create high energy ATP from its lower energy starting materials.

The ATP manufacturing enzyme is closely analagous to a water-wheel, for as the hydrogen ions are allowed to flow back through the enzyme, just as water flows over a water-wheel, so the ensuing chemical reactions 'lift up' the precursors of ATP into their high energy ATP state. So, in a sense, ATP is not a plant cell's primary energy reservoir at all, the hydrogen ion imbalance across the thylakoid membrane is, but this is immediately utilized to generate more permanent energy stores in the form of ATP.

The main overall effects of the solar powered chemical reactions of photosynthesis are quite simple (see figure 6.5). Water, ADP, phosphate and $NADP^+$ are converted into oxygen, ATP and NADPH. This is a chemical reaction just like any other, although more complex than

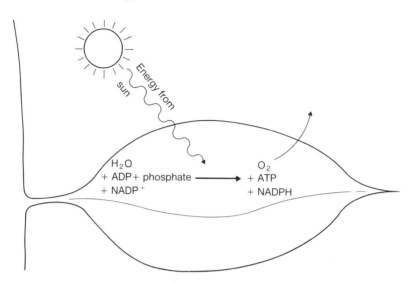

Figure 6.5   The solar powered reaction of photosynthesis in the leaf of a plant.

most, and split up into many separate stages. It is a reaction which raises relatively low energy chemicals up into much higher energy states, so it will only happen if it is driven forward by the automatic dispersal of energy in accordance with the second law of thermodynamics. This energy is provided by sunlight, as it disperses through space from the distant high energy furnace of the sun.

The ATP and NADPH produced by photosynthesis can participate in a variety of reactions with other low energy chemicals of the environment, and, as they do so, they lose some of their own energy and transfer some of it into many of the chemicals of life. For example, ATP and NADPH participate in and power some of the reactions of the carbon fixation cycle, in which carbon atoms from the carbon dioxide of the atmosphere are incorporated into the higher energy carbon-containing compounds of life.

So sunlight lifts chemicals within the cell up into the high energy state of oxygen (which escapes to the environment), ATP and NADPH. The ATP and NADPH (and especially the ATP) then react with other chemicals of the cell to create the high energy chemical complexity of the living world.

Putting the energy to work

ATP is the most important form in which the sun's energy is stored and then used to drive forward the energy-requiring reactions of the living cell. As I have said earlier, many of the most vital chemicals of life are rather high energy chemicals, in which the atoms involved are arranged in configurations which are far more energetic than they could be. The reason living things, or dead plants and bodies, will burn, is because many of their chemicals are in a high energy state. When they burn, the chemicals undergo reactions with oxygen which convert all their carbon atoms into the much lower energy form of carbon dioxide gas (CO_2), all their hydrogen atoms into the lower energy state of water (H_2O), all their nitrogen atoms into lower energy states such as that of nitrogen dioxide (NO_2), all their sulphur atoms into lower energy states such as that of sulphur dioxide (SO_2), and so on.

When cells are being constructed from low energy starting materials, a great many of the chemical reactions involved will proceed only if some energy is available to make the reactions go. Ultimately, that energy is the energy which disperses to the earth from the sun, but sunlight cannot directly power all the thousands of reactions involved. Instead, it powers just one – the reaction which kicks an electron out of a chlorophyll molecule and into an unstable high energy state. Some of that energy is then trapped in the form of ATP and NADPH, and all the other energy-requiring reactions of life are either powered by the breakdown of ATP or NADPH, or by the breakdown of other high energy compounds which have themselves been created by the energy provided by the breakdown of ATP or NADPH.

We can find a good example of the use of ATP as an energy supply for metabolism by reconsidering a fundamental biochemical pathway of life – the one which allows the carbon atoms of carbon dioxide gas to be incorporated into the glyceraldehyde phosphate molecules which form the basic raw material for all the other organic chemicals of life.

In chapter 4 we considered the chemical reactions of the carbon fixation cycle, but no mention was made of the fact that ATP was required to drive some of the reactions forward. You were told that molecules of carbon dioxide react with molecules of ribulose diphosphate to form molecules of phosphoglycerate (see figure 4.3); and then you were told that these molecules of phosphoglycerate were converted by a series of two enzymes into molecules of glyceraldehyde phosphate – a central metabolite used as the carbon- and hydrogen-containing raw material of carbohydrates, proteins, nucleic acids, lipids and much else. The problem I neglected to explore, however, was that the conversion of phosphoglycerate into glyceraldehyde phosphate is an energy-requiring process. It will never proceed unless energy is somehow directly supplied to make it go, or it is somehow coupled to an energy-releasing process, so that the force of the energy-releasing process automatically drives the coupled energy-requiring one forward. This latter mechanism of 'energy coupling' is the one that occurs within the cell. During the formation of glyceraldehyde phosphate from phosphoglycerate, the enzyme which performs the first alteration to the phosphoglycerate – the main energy-requiring step – also catalyses the coupled breakdown of ATP into ADP and phosphate. By 'coupled' I mean that one reaction cannot proceed without the other. As the enzyme breaks down ATP, so it also performs the required conversion on the phosphoglycerate, and it is the force of the breakdown of ATP which actually pushes the phosphoglycerate up into its new high energy configuration.

The principle of such energy coupling can be understood by the simple analogy of the water flowing downhill over a water-wheel, and thus serving to turn the wheel and, for example, raise some weight from the ground using a pulley. If the wheel is linked to the pulley and the weight and put into the stream, then the flow of the stream and the lifting of the weight have become inseparably coupled – one cannot happen without the other. In other words, the water cannot flow past the wheel without causing the weight to be lifted, and the weight cannot be lifted unless water flows past the wheel.

Perhaps a better way to appreciate what actually happens when energy-releasing and energy-requiring reactions are coupled together is to regard a high energy chemical as a compressed spring (see figure 6.6). We saw on page 128 that ATP is a high energy chemical because its structure involves considerable resistance to the electromagnetic force. In particular, the structure of ATP involves three slightly positively

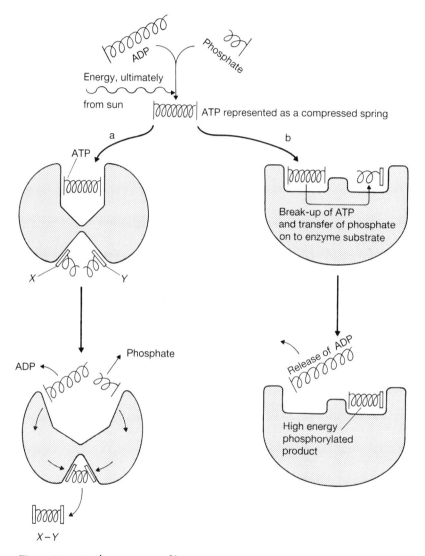

Figure 6.6   A schematic view of how enzymes can couple the release of energy from ATP to the manufacture of other high energy compounds.

charged phosphorus atoms being held close together, setting up an electromagnetic repulsion which strains to make the atoms 'spring apart'. So ATP is really very like a compressed spring – a spring in which electromagnetic repulsion between the phosphorus atoms takes the place of the tension within coiled metal which strains to force a compressed spring back to its uncompressed state.

Figure 6.6a shows you, in a very schematic way, how an enzyme molecule can couple the release of the energy of ATP to the creation of some other high energy chemical. The enzyme can bind to a molecule of ATP, as well as to two low energy starting materials X and Y. The enzyme can then catalyse the break-up of ATP and channel some of the energy released in this process into the creation of the high energy chemical X-Y. In the figure, the release of the tension in the ATP spring is seen to physically force the enzyme into a new conformation in which X and Y are forced together and made to react to form X-Y. The release of the compression of one spring has served to compress another spring. Of course, the compression in the second spring must always be slightly less than the compression of the first spring. In other words, the break-up of an ATP molecule can only power a reaction which requires less energy than the ATP molecule releases. Provided that condition is always met, the enzymes of life can catalyse a wide variety of reactions in which low energy substrates are converted into high energy products. Everything proceeds in accordance with the laws of thermodynamics, provided the energy-requiring reactions are coupled to energy-releasing ones, such as the break-up of ATP, so that in all such coupled reactions a little energy is released into the environment overall.

The energy which is released will, ultimately, have come from the sun, and by releasing some of the sun's energy into the environment the so-called 'energy-requiring' processes of life actually disperse the sun's energy in accordance with the second law of thermodynamics.

The analogy of figure 6.6a is much closer to reality than you might think. We have already seen that the high energy structure of ATP really is very similar to the high energy state of a compressed spring. The way in which that energy can be harnessed by an enzyme and used to create another high energy chemical can also be quite close to our simple physical analogy. The break-up of ATP really can force an enzyme molecule into a new conformation in which the creation of a high energy chemical from its lower energy precursors is encouraged. As ever, all the pushing and pulling is brought about by the electromagnetic force: the release of the electromagnetic tension of the ATP structure can cause new electromagnetic tensions in the structure of an enzyme, causing it to shift its conformation into a structure which forces its substrates to become rearranged into their new chemical form, a form which embodies some of the electromagnetic tension released from the ATP.

A very common alternative mechanism of coupling involves the transfer of the phosphate group released from ATP on to some low energy substrate, or even on to the surface of a protein itself. This process is a bit like transferring part of the ATP spring on to some other molecule, so that this other molecule now contains some of the energy previously held by the ATP (see figure 6.6b). This is what happens when the phosphoglycerate of figure 4.3 is converted into 'diphosphoglycerate'. Diphosphoglycerate carries an extra phosphate group which comes directly from the coupled break-up of ATP. So the break-up of ATP forces the phosphate group released from the ATP to combine with phosphoglycerate, thus creating diphosphoglycerate, which is a higher energy molecule due to its possession of the phosphate group released from the ATP.

Do not be misled by the fact that the break-up of any single ATP molecule can only power a reaction which requires less energy than the ATP releases. This does not mean that only processes requiring less energy than is stored in one ATP molecule can be driven forward in the cell. All it means is that processes requiring much more energy than one ATP molecule contains must be achieved in easy steps, with each small step being driven by the break-up of ATP. Just as a man can lift up a car, by a long series of rather easy pushes on a jack, so the cell can achieve chemical transformations requiring a great deal of energy via a series of small chemical steps, each step requiring less energy than is provided by the break-up of ATP. The energy made available to the cell in the small compressed springs of ATP, can be used to jack up the chemistry of the cell into whatever high energy state is required.

So, in all cases where low energy starting materials are converted into higher energy products, that process is coupled to and driven by the conversion of some high energy chemicals, such as ATP, into lower energy ones. No energy is ever 'given for free' in the cell, and no reactions in the cell ever break thermodynamic laws. Every reaction involving some chemicals being lifted into higher energy levels is accompanied by another reaction in which other chemicals give out at least as much energy, and usually a bit more. There is, of course, one obvious exception – the reaction in which the energy of sunlight kicks an electron out of a chlorophyll molecule up into a high energy excited state. That is the reaction which traps the freely available energy of the sun, allowing it to be stored in a form in which it can be used to drive all the other energy-requiring reactions forward.

The chemical energy of ATP is required to drive forward many of the processes we have considered earlier in this book. Thus, crucial reactions involved in the manufacture of new DNA and RNA and proteins and lipids and carbohydrates would never proceed unless they were somehow coupled to the break-up of ATP. The cycle of muscle contraction is another example. The conformational changes of proteins which underlie muscle contraction (see page 97) are powered by the coupled break-up of ATP.

The details of all the various coupling processes do not matter here. The one simple general principle does: chemical reactions which raise the raw materials of life up into high energy products can proceed within the cell because they are coupled to other reactions which release more energy than the energy-requiring reactions need. The necessary coupling is achieved by enzymes and other proteins when they selectively bind to appropriate high and low energy chemicals and couple the release of energy from one to the incorporation of energy into the other. These proteins are the molecular machines which take the place of the water-wheels and ropes and pulleys which can couple the falling of water down a mountainside to the lifting of some weight beside the stream.

All animals are vegetarians

We have now seen in outline how plants can manage to live and grow and give rise to new plants, despite the fact that to do so many energy-requiring chemical reactions must occur within the plant cells. A very specialized system of membranes, proteins and other chemicals within the plants is able to trap the freely available energy of the sun, and then store it in the form of high energy chemical compounds such as ATP. The breakdown of these high energy compounds into lower energy ones then provides the driving force to power every energy-requiring reaction of the cell, thanks to the chemical coupling of energy-requiring and energy-releasing reactions into combined reactions which always give out energy overall, but in which some of the chemicals involved are pushed up into higher energy states. So much for plants, but what about us? We are unfortunately incapable of utilizing the clever and complex chemistry of photosynthesis, so how do high energy chemicals such as proteins and nucleic acids and lipids and carbohydrates ever get created within us? The answer lies with the principle of chemical coupling, and the ability of animals either to eat plants, or eat other animals which have eaten plants.

The principle of chemical coupling suggests that animals will be able to live and grow if they can somehow manufacture ATP, since that ATP could then be used to drive forward all the energy-requiring reactions of animal life; but how can we make ATP if we are incapable of photosynthesis? Well, we can eat the corpses of plants, or other animals which have themselves been sustained by eating plants, and we can convert the high energy chemicals in these corpses into low energy ones, and couple that energy-releasing process to the energy-requiring manufacture of ATP.

Suppose you have just had a cup of black coffee with sugar. The drink will have contained some sustenance for you in the form of the sugar molecules which were initially manufactured within the cells of a plant. What happens to the high energy sugar molecules to provide you with ATP? We could easily confirm that the sugar molecules really are in a high energy state by burning them in air. They would react with the oxygen of the air to generate carbon dioxide and water, and a lot of energy in the form of heat. That same overall reaction is essentially the way in which the energy of the sugar is made available to your body, some of it being released as heat, but some of it being trapped in the chemical form of ATP.

One of the most common and most central types of sugar molecule is 'glucose' (common 'sugar' consists of glucose linked to another sugar known as 'fructose'), and by looking at the way in which our cells can capture much of the energy of glucose we will see all the general principles of how animals get their energy.

What happens overall is that the atoms of a glucose molecule are combined with atoms of the oxygen we breathe in from the air to generate carbon dioxide and water as wastes; but in animal cells this simple overall chemical reaction occurs in a very complex and indirect way. The main features are shown in figure 6.7. It shows that first of all, in the cytosol of a cell, glucose molecules are converted, by the activity of many different enzymes, into molecules of a simple three-carbon compound known as 'pyruvate'. The pyruvate molecules then enter the specialized organelle known as the mitochondrion, where they are fed into a complex metabolic cycle known as the 'tricarboxylic acid cycle' (or sometimes the 'citric acid cycle' or the 'Kreb's cycle'). During the operation of this cycle carbon, hydrogen and oxygen atoms derived from glucose are continuously fed into the cycle, and by the time each cycle is complete these atoms have been converted into the form of carbon dioxide and water.

The details of this 'oxidation' of glucose need not concern us, for the overall principles are what we are interested in. What is happening overall is that the atoms of glucose are effectively reacting with oxygen to generate carbon dioxide and water. That overall reaction releases a great deal of energy, and much of this energy is captured, using the principle of chemical coupling, by being used to manufacture ATP. Again, the details of the chemical coupling reactions are complex, but they work according to the simple principles examined above.

So animals are able to capture some of the energy of the sun, not directly, but indirectly by eating plants and allowing the high energy chemicals of the plants to react with oxygen to reach low energy states, a process which drives forward the manufacture of the ATP which can be used as a primary energy source by the animal cells.

Notice that animals get their energy by reversing the processes by which plants use the sun's energy to create sugars (such as glyceraldehyde phosphate) and other high energy compounds (see figure 6.8). In plants, the sun's energy is harnessed to convert water and carbon dioxide into sugars and oxygen; in animals, sugars are reacted with

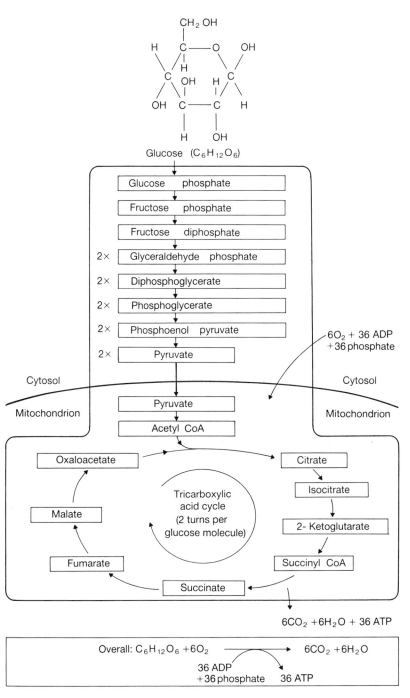

Figure 6.7   The oxidation of glucose.

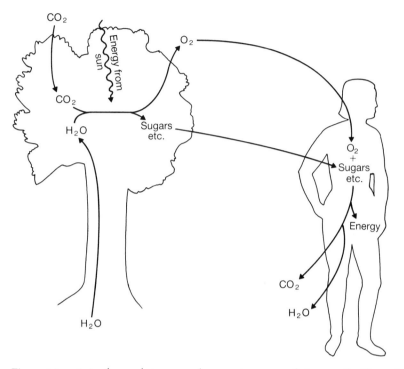

Figure 6.8   Animals get their energy by reversing some of the overall effects of photosynthesis.

oxygen to generate carbon dioxide and water, accompanied by the manufacture of high energy ATP. In fact, the second process, the oxidation of sugars, is also performed by plants when they wish to convert some of their energy stores (largely held in the form of complex carbohydrates) back into ATP.

Animals are completely dependent on plants for their supplies of energy, because only plants (and various photosynthetic bacteria) are able to capture the sun's energy and thus act as a primary energy source for all other life. An animal which eats another animal is really getting its energy from plants, because the eaten animal will itself have relied on plant foods to build up its body, or it, or some smaller animals farther down the 'food chain', will have eaten other animals which relied directly on plants. A long food chain may exist, composed of animals which eat other animals, which eat other animals, and so on; but eventually, at the bottom of the food chain, there will be animals who are sustained directly by eating plants. In this sense all animals are vegetarians, because the energy their food supplies to them has originally been captured from the sun by the process of photosynthesis in plants (or in photosynthetic bacteria). Also, almost all the atoms in animals have originally been taken up from the environment by plants.

Fermentation and other odds and ends

Some simple forms of life, such as yeasts and some bacteria, cannot make their own high energy chemicals by photosynthesis, but neither can they extract energy from high energy foods by combining them with oxygen. Many of these creatures (usually single-celled) can survive and grow in the complete absence of oxygen, indeed some of them are killed by oxygen. So how do they make ATP?

In many cases, such as yeasts, they make ATP by performing only the first stage of the reaction scheme shown in figure 6.7 – the stage which proceeds in the cytosol and does not require oxygen. Yeast cells can convert glucose, for example, into pyruvate molecules, and then convert the pyruvate into 'ethanol' (i.e. alcohol) and carbon dioxide wastes. The ethanol and carbon dioxide contain the atoms of glucose in a slightly lower energy configuration, and some of the energy difference is trapped, via a simple chemical coupling reaction, in the form of ATP. Processes such as this, in which a little of the energy available in high energy chemicals such as glucose is used to make ATP without the involvement of oxygen, are known as 'fermentations'. Many simple organisms other than yeasts can make ATP by fermentative processes which all convert high energy chemicals into slightly lower energy wastes (not always ethanol and carbon dioxide), but which leave much of the energy available in their foodstuffs untapped.

A few types of bacteria can perform photosynthetic reactions which use hydrogen sulphide (H_2S) in place of the water used by plants; while some cells can extract chemical energy by combining high energy inorganic chemicals with oxygen to produce lower energy wastes. These last types of cells are the only ones on earth which can live and grow without relying, either directly or indirectly, on the energy of sunlight.

The vast majority of species on earth are dependent on the energy of the sun to power the creation of the high energy chemicals they need to live and grow. Plants and photosynthetic bacteria can trap the sun's rays directly, while most other life-forms, such as ourselves, must extract some of the energy from foodstuffs whose energy can ultimately be traced back to the plants, and therefore to the nuclear fusion reactions within the sun.

You are solar powered, and so is just about every other type of creature on earth.

7   Keeping control

Cells are often likened to factories, with their various metabolic pathways being likened to production lines. Raw materials enter through various 'doors' in the membrane. Processing production lines convert the raw materials into the required starting materials for the main manufacturing production lines, which then assemble the components needed by the factory. Other production lines are constantly breaking down many of the factory's products, and also the very fabric and machinery of the factory itself, to maintain the metabolic 'turnover' which facilitates growth and flexibility.

Of course some of the raw materials are not used for the manufacturing production lines, but are degraded directly on lines which produce the energy (in the form of ATP molecules) needed to power all the other processes on the factory floor. The inward flow of raw materials is accompanied by a steady outward flow of simple waste products which must be returned to the environment.

The production line analogy is a very good one. The individual 'work stations' are the enzymes, and at these molecular work stations various chemical components are brought together and fashioned into some new component or product. The product of one enzyme can then pass on down the line, to become the substrate of the next enzyme, and so on until the pathway is complete.

This analogy also reveals one of the great problems which the chemistry of the cell must somehow overcome – the problem of the co-ordination of all the cell's activities, the problem of control, the problem of integrating the chemical potential of the living cell into a smoothly operating whole. Consider what would happen in a factory in which there was no means of controlling the various production lines or what was happening at the various work stations. Suppose one line started going too fast. Its products would be churned out endlessly at the end of the line to accumulate in a massive pile which might grow uncontrollably until it swamped the entire factory and prevented all the other production lines from working. Suppose some essential component began to be used up. If there was no means of telling the other production lines to make more of the component, or of gathering it up more quickly from the environment, then first one line, and eventually the entire factory, might grind to a halt. So it is no good simply having the ability to perform all the various processes required to generate a living cell. These processes must all proceed at appropriate speeds, and mesh together in appropriate ways if an ordered cell is to be created rather than the messy chemical chaos of death.

If we contend that all the activities of a living cell are the results of 'mere chemistry', then somehow mere chemistry must automatically ensure that the chemical activities of the cell are controlled and co-ordinated in appropriate ways. How can this possibly be achieved?

The central dynamic processes of living cells are their metabolic pathways, in which chemicals are acted upon by a sequence of enzymes which converts them into some products needed by the cell. The enzymes may not be physically arranged in any specific sequence (although in some cases they probably are), but they act upon the chemicals flowing along the pathway in sequence because each enzyme can selectively bind to only the chemicals it acts upon. In many cases the random bumping and dancing of all the molecules of the cell cytosol can be relied upon to deliver each chemical to its appropriate enzymes often enough for the chemistry of life to proceed.

So let us examine a very simple and imaginary metabolic pathway, to see what mechanisms are available for the pathway to be controlled and integrated with the activity of other pathways in a manner consistent with the needs of the cell. The pathway consists of a series of enzymes, each of which converts some substrate(s) into some product(s) (see figure 7.1). The substrates and products are represented in figure 7.1 by boxes labelled with a letter, but each box may represent several different chemicals since many enzyme reactions involve two

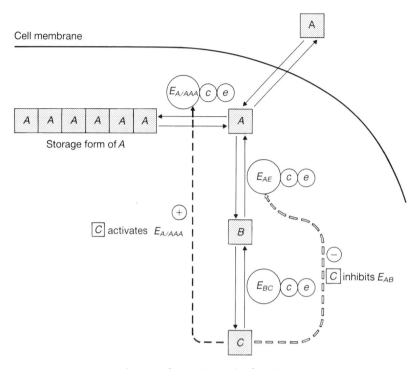

Figure 7.1   Some mechanisms for regulating the flow through metabolic pathways: mass action effects due to the reversible nature of enzyme catalysis; changes in the availability of enzymes, or of cofactors; changes in the intracellular environment; the inhibition or activition of key enzymes by key metabolites (see text for details).

or more substrates coming together on the enzyme to produce two or more products. Thus, some enzymes may convert x and y into z, or into xy, or into zwf; or may split zwf into z and wf, or into z, w and f, and so on.

Each enzyme in figure 7.1 is represented by the letter E, followed by a subscript indicating which substrates it acts upon and which products it creates. Many enzyme reactions also require the presence of small chemical cofactors, to allow the reaction to proceed, and these are represented as c, although not all enzymes depend on such assistance. Many enzymes are also greatly affected by the precise chemical environment around them. An increase in the acidity of the environment, for example, may cause their activity to increase (i.e. make them catalyse their reaction faster) or decrease, or become altered in more subtle ways (such as changing their specificity so that they act on a slightly different set of substrates. Other factors in the environment (represented as e) which may affect an enzyme's activity are the concentrations of various metal ions, or the temperature of the surroundings.

In the figure, the conversion between substrates and products is represented by a double-headed arrow because enzymes generally catalyse *equilibrium* reactions, in which, for example, B can be converted back to A by enzyme E_{AB}, as well as A being converted into B. The net flow of material through this step of the pathway is the result of some imbalance in the rates at which the forward and reverse reactions are catalysed by the overall activity of all the available molecules of the enzyme. The same goes for the net flow of material through any step of any metabolic pathway. With many molecules of any one enzyme available in the cell, the reaction catalysed by these enzyme molecules can be proceeding forwards and backwards at the same time. In many cases the forward reaction may proceed a great deal faster than the reverse one, so virtually none of the product may be converted back into the substrate; but even in such cases it is important to remember than the reactions catalysed by most enzymes are reversible and therefore capable of settling down into some equilibrium state in which the overall flow in any one direction is the net result of the rate at which the forward and reverse reactions are proceeding.

This reversibility of enzyme reactions makes possible the simplest form of automatic metabolic control, in which the availability of starting materials can speed up the flow along a pathway, while large stocks of the product can slow it down. The rate at which any chemical reaction proceeds is directly dependent on the availability of the reactants concerned, for a very simple reason. If there is lots of A around in the cell, for example, then the rate at which A happens to meet up with enzyme E_{AB} will obviously increase (just as an increase in the number of people you happen to know entering a fairground will increase the chances of you meeting up with someone you know). Since E_{AB} will meet up with more A, it will convert more of A into B in any given time, or, in other words, the rate of the reaction converting A into B will increase. This effect, in which increasing the 'concentration' (the amount in a given volume) of a chemical inevitably increases the rate of any reactions it participates in, is called the law of 'mass action' (since increasing the mass of the chemical available increases the rate at which it reacts). Thanks to the law of mass action, and the fact that enzymes catalyse reversible reactions, the rate of flow along a metabolic pathway can automatically adjust itself to fit in with the availability of the starting materials or the accumulation of products.

Suppose, for example, that a massively increased supply of A suddenly becomes available to the cell. If we were dealing with a factory production line someone would need to activate some switch or lever to make the line move faster to exploit the increased availability of A. A metabolic pathway, however, will speed up automatically thanks to the law of mass action. Since much more of A is available in the cell, molecules of A will meet up with molecules of E_{AB} much more often, and so *inevitably and automatically* the rate at which A is converted into B will increase. This, of course, will increase the concentration of B, causing the rate of the conversion of B into C to increase as well (and in principle this effect could continue down the entire length of a long metabolic pathway).

So an increase in the availability of a basic starting material can automatically cause a metabolic pathway to 'speed up'. Eventually, however, the extra supplies of the starting material will become used up and the pathway will automatically slow, simply because less starting materials will then be available. There is another way in which the law of mass action can cause the pathway to slow. Suppose that the increased availability of A causes the simple pathway of figure 7.1 to speed up to such an extent that massive quantities of C begin to build up. C, of course, is not only the product of enzyme E_{BC}, it is also the *starting material* of the reverse reaction. So as the concentration of C builds up, the reverse reaction will accelerate and so the *overall* rate of production of C will decrease. This reverse effect of the law of mass action could be transmitted all the way back along the pathway, to result in a net slowing down of the rate at which A is used up.

So the rate of metabolic pathways can automatically become adjusted in rather simple ways to fit in with the changing availability of raw materials and the build-up of products. An increased supply of raw materials will initially cause a pathway to speed up, so that the supply is made use of. If this eventually causes more of the product of the pathway to be manufactured than the cell can use, then the net flow along the pathway will slow down again as it begins to be driven partially in reverse. This neatly prevents the build-up of the product of the pathway from causing any problems, but of course it does recreate the possible problem of too great a concentration of raw materials. In many cases a continued build-up of raw materials will either cause these raw materials to begin being channelled off along other pathways, such as ones that convert them into some storage form; or the build-up of raw materials will cause the entry of further raw materials to cease, until the supplies of available raw materials have been depleted back to acceptable levels. Both such responses – the channelling off of raw materials along storage pathways, or the shut-down of the entry of the raw materials, could be triggered by mass action effects, but there are many other more subtle ways in which the automatic regulation of metabolism can be maintained, as we shall see later.

If the problem faced by the pathway, however, is not that A has become overabundant, but that C is being used too quickly, then mass action effects can again help out because the rate of conversion of C into B will dramatically reduce, causing an increase in the net conversion of B into C, causing the rate at which B is converted back into A to decrease, causing an increase in the net conversion of A into B, causing the levels of A to decrease, triggering either the increased uptake of A from the environment (if available) or the increased degradation of the storage form of A into free A.

The law of mass action describes the simplest way in which the metabolic pathways of a cell can react to fluctuations in supply and demand: increased supply will cause increased flow along the metabolic pathways which make use of that supply; reduced supply will slow the relevant pathways down; increased demand for the products of the pathways will cause acceleration of the pathways, and may in turn draw in increased supplies from outside or cause an increase in the rate at which stores are degraded into direct supplies. Decreased demand for the products of the pathways can slow the pathways down, and perhaps ultimately stop supplies being taken up from the environment, or cause them to be converted into storage forms.

The law of mass action is an effective way in which the rate of flow along metabolic pathways can react appropriately to fluctuations in supply and demand; but it is not a very efficient one, especially when long pathways are involved. A more efficient system would allow the key points in a metabolic pathway to affect one another directly, rather than by passing 'the message' backwards or forwards through every step of the way. There are various very effective ways in which the key points of pathways (their starts, ends, intermediate branch points, etc. . . .) can 'communicate' directly, and they largely depend on various mechanisms of enzyme 'activation' and 'inhibition'.

A good, simple and very common example is the inhibition of an early enzyme in a pathway by the end-product of that pathway – a situation known as 'feedback (end product) inhibition'. Suppose that one of the products of the enzyme E_{BC}, something in the C box, in other words, is able to bind to the surface of enzyme E_{AB} and bring about a change in the enzyme conformation which greatly reduces its activity. The end product of the pathway will be an inhibitor of the first step in the pathway, an ability which would prevent the pathway from overproducing C. A great many metabolites are able to act as enzyme inhibitors in this way. They can inhibit enzymes of both the pathway which generates them, and also of other pathways. So the accumulation of specific metabolites in the cell can cause particular metabolic pathways to be inhibited and therefore slowed down.

The reverse situation also occurs, in which specific metabolites are able to bind to particular enzymes and *activate* them, so that they will then perform their acts of catalysis much faster. In our imaginary example of figure 7.1, it would be sensible if some component of C were able to bind to and activate a crucial enzyme of the pathway which converts A into some storage form. This would ensure that at times when A was in plentiful supply, and plenty of C was being produced, the excess supplies of A could be stored for times when A becomes scarce.

Most of the signalling tasks involved in regulating and co-ordinating the various pathways of metabolism depend on crucial metabolites which can either inhibit or activate selected enzymes. The metabolic pathways of the cell are co-ordinated into a meaningful and efficient whole by many complex networks of such inhibition and activation.

One other crucial activity of the cell which must be subject to careful control is the activity of the genes which encode the enzymes and other proteins of the cell. Genes can be 'switched on or off' by interactions between the DNA of a gene and some protein, or some metabolite, or some RNA, or some combination of these agents.

For example, all cells make various 'gene regulating proteins' which are able to bind to specific regions of DNA close to genes and switch the nearby genes on or off, depending on the proteins concerned (see figure 7.2). Some of these proteins prevent nearby genes from being used to manufacture mRNA, by physically preventing the enzymes responsible for mRNA manufacture from gaining access to the DNA of a gene. Such 'repressor proteins' can be controlled by small metabolites which bind to them and alter their conformation in a way which makes the protein no longer able to bind to DNA. So appropriate metabolites can switch genes on – 'activate' them – by binding to appropriate repressor proteins.

Conversely, cells also contain gene 'activator' proteins, which are able to bind to specific regions of DNA and activate nearby genes so that they begin to be copied into mRNA. Such activator proteins can be controlled by small metabolites able to bind to an activator protein and alter its conformation in a way which makes it no longer able to bind to DNA. So appropriate metabolites can switch genes off by binding to appropriate activator proteins.

In some cases activator proteins can only activate genes when the activator proteins themselves are activated by the binding of some metabolite, while some repressor proteins can only repress genes when the repressor proteins are made active by the binding of some metabolite. So there is a variety of ways in which the levels of various metabolites in a cell can influence the activity of the cell's genes.

Some gene-regulating proteins are able to activate or repress a wide variety of different genes simultaneously. In other words, many copies of such proteins can become bound to many different sites on the DNA of a cell to cause some large-scale co-ordinated change in the battery of proteins available in the cell. Some proteins are able to bind to and activate or repress the gene-regulating proteins themselves.

The decoding of genetic information into working protein molecules can also be controlled at later stages in the process other than the

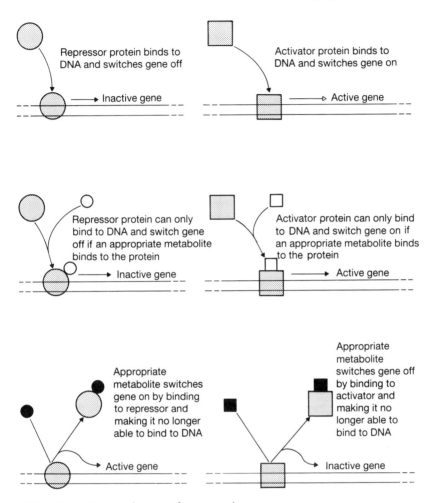

Figure 7.2   Some mechanisms of gene control.

initial production of mRNA. The 'processing' of the mRNA, in which the sections which actually code for proteins are spliced together while the introns are cut out, can be a site of control. The transport of mRNA from the nucleus to the cytoplasm can be subject to control. The chemical reactions of the translation of mRNA into a protein can be subject to control. The rate at which mRNAs become degraded and therefore inactive can be subject to control.

During the process of development, in which a fertilized egg, for example, divides and develops into a large multicellular organism, all the cell divisions and cell changes must be controlled by genes, proteins and metabolites interacting with one another and with their environment. The net effect of this network of regulatory interactions is to ensure that the right genes become active in the right cells at the right times. It ensures that the genes required to make brain cells are active only in brain cells; that the genes required to make muscle cells are active only in muscle cells, and so on.

The *details* of all the required regulatory interactions between genes, proteins, metabolites and their environment are unknown, and form one of the most active areas of currrent biological research. Indeed, in many cases the details of how all the various methods of control outlined in this chapter interact in appropriate ways are still unknown. The *principles*, however, seem secure, and this is a book of basic principles rather than details. The fundamental principle underlying the integration and co-ordination of the chemistry of the cell is that some chemicals can alter the activity of other chemicals. That ability gives the 'factory' of the cell an effective means of passing signals between its various 'production lines', allowing all the lines to work at appropriate speeds and at appropriate times to promote the survival and growth of the cell.

8 The nervous mind

At the heart of our experience of life there is a deep and subtle paradox. Most of us consider the world of physical matter to be the trustworthy world of true reality. We think of rocks and rivers and tables and chairs and our own flesh and bones as objects composed of solid certainty. The world of consciousness and thought, on the other hand, is regarded with more suspicion as an abstract and ghostly place in which there are no certainties and great doubt about the true nature of its reality. The physical world is the world we feel comfortable with, while the world of the mind seems a domain of mystery and even magic. The paradox is that the world of the mind is the only one we have direct experience of, and the only one each of us can be certain exists. Our experience of the physical world comes to us indirectly, along nerves which convert the effects of that physical world into sensations, thoughts and ideas in our brain. Each one of us can be certain that our own thoughts exist, but that is the only thing we can be absolutely certain about.

Since our conscious mind is the part of our experience of life which really matters, indeed it is all that we really are, we surely cannot pretend to understand the inner mechanisms of life until we understand the workings of the mind. Sadly, the mind remains the hidden inner chamber of biology whose secrets are largely unknown and unexplored. We know a lot about the nerve cells upon whose activity the mind apparently relies, and we are constantly learning more. This chapter will tell you the most fundamental facts and principles about what these nerve cells seem to do. Yet at the moment we can only look at that body of facts and say 'somehow the overall effect of it all is the creation of a conscious mind.' There is a great assumption on the part of biologists that the nature of a conscious mind really is as simple as that – that it arises 'somehow' from the integrated activities of the chemicals within the brain. Others contend that the origin of the mind lies in aphysical 'spiritual' or 'supernatural' phenomena which merely rely on the physical brain to provide a suitable habitat in which the mind can reside and be sustained. Nobody has any satisfactory suggestions as to what the nature and origin of that aphysical essence might be.

My claim to be able to reveal to you the secrets of life may be fraudulent, if it is expected to extend to conscious, reasoning life-forms rather than unthinking (presumably) single cells. Many scientists, however, believe that the three secrets revealed in the Introduction to this book really do cover the origin of conscious minds as well as of living cells. They believe that, in essence, there really is not 'anything more' to a conscious mind than the overall effects of the electromagnetic force interacting with energy to make chemical reactions proceed, the ability of nucleic acids to direct both their own reproduction and also the production of proteins, and the ability of proteins to directly or indirectly promote the chemical reactions that create and sustain life, and therefore minds.

Let us go along with that view for a while, to investigate the physical mechanisms which have been uncovered at work within the nervous tissue which lets us think.

Nerves

The fundamental structure within the nervous system is a highly complex and organized network of interconnected nerve cells. Various other types of 'supporting' cells are present as well, but the nerve cells are believed to be the ones that really matter. In general terms nerve cells are simply cells like any other cells. They are bounded by a lipid bilayer membrane which has proteins embedded in it. They contain a nucleus and mitochondria and all the other chemical paraphernalia of the living cell. What, then, is special about them? The answer is that they are specialized to make contact with one another and affect one another's activities in a bewildering complexity of ways. We call all this contact and its effects 'nervous communication' and say it is mediated by nerve 'impulses', which are simply electrochemical signals able to pass along and between nerve cells. When the mechanisms of nervous communication are examined in detail there appears to be nothing mysterious or magical or particularly astonishing about it at all. Like most of the events of life, it is brought about by the direct and indirect activity of proteins due to their ability to perform feats of selective chemical binding, catalysis and conformational change.

Nerve cells vary greatly in their shape and structure, but they all adhere to the same basic plan (see figure 8.1). They have a 'cell body'

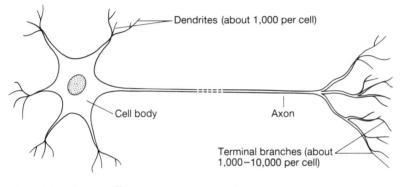

Figure 8.1   A nerve cell.

which houses the nucleus and some surrounding cytosol containing all the usual organelles and metabolic processes. Many spidery extensions of the cell body form thin 'limbs' known as 'dendrites'. The dendrites are the main parts of the cell which receive impulses from other cells. Each nerve cell also has a longer extension known as the 'axon' which in some nerve cells can be anything from 1 millimetre to an astonishing 1 metre in length. Close to its end, the axon branches into many thin limbs known as the 'terminal branches' of the axon. The ends of these branches are the parts of the cell that pass impulses on to other cells. The entire structure, consisting of nerve body, dendrites, axon and terminal branches, is surrounded by the cell membrane, and even the finest of the branches contains a small amount of cytosol.

To explore what a nerve impulse is, we must consider more closely the membrane of a nerve cell and the specialized proteins that lie embedded in it. The main proteins which matter are ones which can facilitate and control the passage of small ions through the membrane, such as sodium (Na^+) ions and potassium (K^+) ions. The cell membrane is normally impermeable to all types of ions. Ions carry electrical charge and so it would take a lot of energy to make them move out of a watery region, where they can interact with the slight charges carried by water molecules, and enter the uncharged hydrophobic interior of the membrane where they would disrupt the Van der Waals interactions between the molecules of the membrane. Since it would take a lot of energy for an ion to move into and through a membrane, it does not happen. Ions can move between the outside and the inside of a cell, however, if proteins help them to do so. Proteins can do this by forming trans-membrane channels through which ions (often only certain specific ions) can diffuse freely; or proteins can actively bind to specific ions and then undergo a conformational change which whisks the ions across the membrane.

The most important ion-transporting protein of all cells, not only nerve cells, is one which can transport Na^+ ions out of the cell at the same time as transporting K^+ ions in (see figure 8.2). The activity of this 'Na^+/K^+ pump' is coupled to the break-up of ATP, allowing it to pump Na^+ ions out until their concentration outside is much greater than inside the cell, and to pump K^+ ions in until they are far more concentrated inside than out.

Another membrane protein forms a channel through which K^+ ions can passively diffuse much more effectively than Na^+ ions. So this second protein forms a selective 'K^+ leak channel'. K^+ ions will automatically pass outwards through this channel much more frequently than they pass inwards, simply because there are many more K^+ ions inside the cell than outside due to the activity of the Na^+/K^+ pump. The preferential outwards movement of K^+ ions makes the layer of solution just outside the cell positively charged relative to the layer of cytosol just inside the cell. This charge imbalance builds up until it reaches the level at which the tendency of K^+ ions to leak outwards is exactly balanced by their tendency to be pulled back in by the electromagnetic attraction between the positive K^+ ions and the

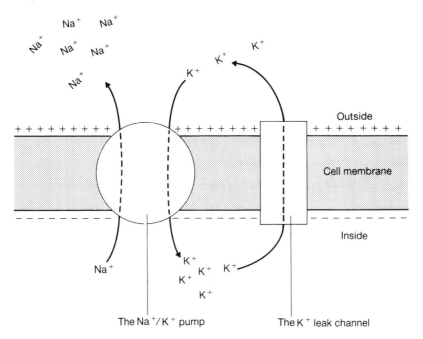

Figure 8.2   Cell membranes are electrically polarized, being positively charged on the outside and negatively charged on the inside, due to the action of the Na⁺/K⁺ pump and the K⁺ leak channel.

negatively charged interior of the cell. So, in another nice example of the automatic self-regulating powers of chemistry, the build-up of a charge imbalance across the membrane automatically stops itself from building up further once a certain level is reached.

All the time, of course, the Na^+/K^+ pump is pumping K^+ ions back into the cell, and so allowing other K^+ ions to leak out in compensation. If the charge imbalance ever drops a little below its equilibrium level, more K^+ ions will automatically leak out to restore it; while, if it ever becomes fractionally too large, the outward movement of K^+ will cease, or even reverse, until the charge imbalance returns to its equilibrium point.

So the combined effect of the Na^+/K^+ pump and the K^+ leak channel is to maintain a slight charge imbalance across the cell membrane, at the same time as keeping the K^+ ion concentration far higher inside the cell than outside, and the Na^+ ion concentration far higher outside than inside.

This charge imbalance exists across the membrane of all cells, and it is described using various terms. Obviously it means that there is an electrical 'voltage difference' across the membrane. Another name for such a voltage difference is a 'potential difference', and so the charge imbalance across a cell membrane is sometimes referred to as the 'membrane potential'. Another way of describing the situation is to say that the membrane is 'polarized'.

The movement of other types of ions, such as Ca^{2+} and Cl^- ions, through the membrane (thanks to appropriate transport proteins) also plays a part in determining the overall electrochemical polarization of a cell membrane. These other ions, however, play more minor roles. The central players in the generation of the polarized membranes of all cells are the Na^+/K^+ pump and K^+ leak channel proteins, and the Na^+ and K^+ ions whose movement these proteins facilitate.

The membranes of all cells are polarized in this, or a similar, way, but nerve cells are able to make special use of the situation thanks to some other types of protein which are found embedded in the membranes of nerve cells. The 'special use' nerve cells make of polarized membranes is to allow them to generate and transmit nerve impulses. We still have not discussed what a nerve impulse actually is, but do not worry, we are getting closer!

Let us consider the basic principles involved in the creation and transmission of a nerve impulse within and between nerve cells. When a nerve cell is induced to transmit a nerve impulse along its length, and then pass it on to any other cells it is connected to, we say that the nerve has been induced to 'fire'. What causes a nerve cell to fire is the arrival of a chemical signal which will have been released from another cell (or, more usually, the overall effects of many different chemical signals arriving from many different cells – see page 167).

Imagine three connected nerve cells A, B and C, with the terminal branches of A connected to the dendrites of B, and the terminal branches of B connected to the dendrites of C. Every now and then, for reasons which will become clear shortly, the terminal branches of cell A will begin to release molecules known as 'neurotransmitters' into the narrow gaps between the terminal branches of cell A and the dendrites of cell B. These narrow gaps are known as 'synapses'. The membrane of cell B will contain receptor proteins which are able to bind to the neurotransmitter molecules, and then undergo a very significant conformational change in response to that binding. The conformational change will cause a channel to appear through each of the protein molecules, which will allow ions to diffuse freely in and out of cell B. The binding of the neurotransmitter molecule to its receptor protein effectively punches a tiny hole through the membrane, although passage through this 'hole' is often restricted to only certain types of ions. Positive ions (often Na^+ ions) will be able to flood into the cell, pulled by the electromagnetic force which attracts them to the relatively negatively charged interior, and so the membrane around the receptor protein will quickly become 'depolarized'. The ion imbalance which created the polarization will be destroyed.

Another vital type of protein found embedded in the nerve cell membrane is one which is caused to change between two conformations when the membrane in which it is embedded becomes depolarized. When it is sitting in a polarized membrane, this protein is in a conformational state in which it is unable to allow any ions to pass through the cell. When the membrane around it becomes depolarized, however, the protein undergoes a conformational change which causes it briefly to form a channel through which Na^+ ions can pass. The channel only remains open for a short time, however, since the conformational upheaval of the protein continues until it adopts a new conformation in which the passage of Na^+ ions is once again blocked. The overall effect of this conformational change is a bit like the operation of a turnstile – it moves from one conformation which prevents anything from passing, into a new conformation which also prevents anything from passing, but in the process of changing from one conformation to another there is a brief period during which a channel allowing passage though it is opened up.

Of course while this Na^+ channel is open Na^+ ions will flood into the cell, since they are still at a very high concentration outside and at a very low concentration inside. The effect of this entry of Na^+ ions is to depolarize a larger region of the cell membrane all around, and then it proceeds to cause a significant *reverse* polarization to build up, with the interior of the cell becoming positively charged relative to the outside.

Molecules of the 'voltage-sensitive Na⁺ channel' are studded at regular intervals throughout the membrane of the nerve cell. Some of them will be near to the receptor proteins which depolarize a small region of the membrane around them when they bind to an appropriate neurotransmitter. This initial membrane depolarization will cause a few of the neighbouring Na⁺ channels to undergo their conformational change; and then the polarization reversal caused by the effects of the first Na⁺ channels to open will induce neighbouring Na⁺ channels to undergo the same change. In this way a *wave* of polarization reversal can spread rapidly along the entire length of a nerve cell. This is what happens when we say that a nerve cell has been induced to fire. We mean that it has been induced to transmit a wave of polarization reversal across the entire cell membrane, from one end of

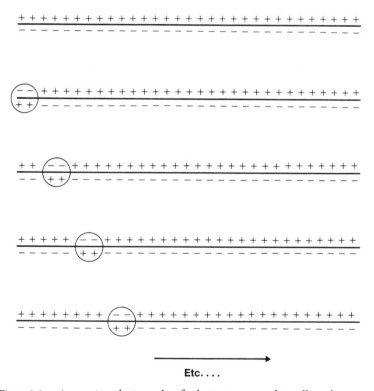

Etc. . . .

Figure 8.3   A nerve impulse is a pulse of polarization reversal travelling along a nerve cell membrane. It is created by the sequential opening and closing of protein ion channels which control the passage of ions through the membrane.

the cell to the other (see figure 8.3). This spreading wave of polarization reversal is a nerve impulse. It also goes by the name of an 'action potential'.

A nerve impulse really does spread along a nerve cell as a discrete 'pulse', with regions of normal polarization both ahead of it and behind it, because, as one region of the cell is undergoing its polarization reversal, earlier regions are recovering their normal polarization. This recovery occurs because, first, the neurotransmitter molecules which caused the nerve cell to fire soon dissociate from the receptor protein and become degraded (by enzymes) so the receptors quickly revert to their closed state; and secondly, the Na^+ channels which opened up briefly during their induced conformational change soon settle into their new closed conformation. So, shortly after the wave of polarization reversal has passed, things are returning to normal and the normal membrane polarization is being set up and maintained as before. In many cases the return to normality is assisted by the activity of various other membrane proteins, some of which become able to let K^+ ions flow out of the cell for example. The Na^+ channels also slowly return to their *original* closed conformation, via a different conformational change which does not involve any opening up of their channels.

So a nerve impulse is a wave of electrochemical change which spreads quickly across a nerve cell membrane. It begins when neurotransmitter molecules (made by an enzyme, of course) are released from one cell (or several cells) and become selectively bound to special receptor proteins on the next cell. These receptors undergo a conformational change which causes a depolarization of the membrane around them. This depolarization induces other neighbouring protein molecules to undergo a conformational change which depolarizes and then reverses the polarization of a further area of the membrane, which causes neighbouring protein molecules to bring about further polarization reversal in the same way; and so the process proceeds along the entire length of the nerve cell. The mechanisms at work in the creation and transmission of a nerve impulse are the same familiar automatic mechanisms we have met repeatedly before: the ability of proteins to perform catalysis, selective binding and be induced to undergo conformational change; and the ability of the electromagnetic force to bring about all the chemical reactions and interactions which underly that catalysis, selective binding and conformational change.

But what happens when the nerve impulse reaches the far end of nerve cell *B*, the end containing the terminal branches which must somehow transmit the impulse to other cells such as cell *C*?

The cytosol in the terminal branches of a nerve cell contains many tiny membrane-bound vesicles full of neurotransmitter molecules. The cell membrane in this region also contains proteins which can be induced to open up to form a channel through which calcium (Ca^{2+}) ions can pass when the membrane undergoes depolarization. So the arrival of a nerve impulse at a terminal branch causes these 'Ca^{2+} channels' to open up, this causes Ca^{2+} ions to flood into the cell (since there is normally much more Ca^{2+} outside than inside), and this, in turn, encourages the membranes of the vesicles packed with neurotransmitter to fuse with the cell membrane and thus release their neurotransmitter out into the space (the 'synaptic cleft') between one nerve cell and the next. So the arrival of a nerve impulse at a terminal branch ensures that neurotransmitter molecules are released into the synaptic cleft, allowing them to diffuse across the cleft, bind to the appropriate receptor proteins in the membrane of the next nerve cell, and induce that nerve cell to fire in the way that we have just examined.

These are the bare essentials underlying communication between the cells of the nervous system: neurotransmitter molecules bind to receptor proteins, induce them to change conformation, causing alterations to the internal chemistry of the cell which can induce further conformational changes in various membrane proteins, which can cause a nerve impulse to pass along the cell and bring about the release of further neurotransmitter molecules from the cell's terminal branches (see figure 8.4).

As you might expect, however, there are many additional subtleties and complications. First, there are many different types of neurotransmitter, capable of causing subtly different changes within the cells that they bind to. Secondly, while many 'excitatory' neurotransmitters (like the one in our example) *encourage* nerve cells to fire, other 'inhibitory' neurotransmitters *discourage* cell firing by resisting the depolarization of the membranes of the cells they bind to. Thirdly, most nerve cells receive inputs, in the form of neurotransmitters, from

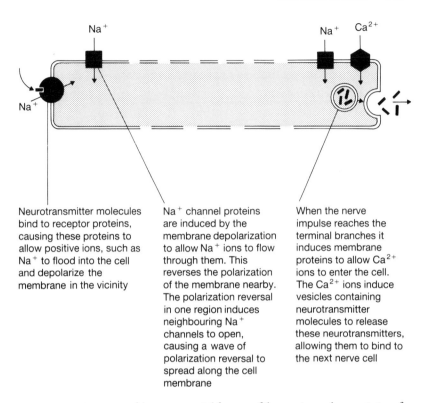

| Neurotransmitter molecules bind to receptor proteins, causing these proteins to allow positive ions, such as Na⁺ to flood into the cell and depolarize the membrane in the vicinity | Na⁺ channel proteins are induced by the membrane depolarization to allow Na⁺ ions to flow through them. This reverses the polarization of the membrane nearby. The polarization reversal in one region induces neighbouring Na⁺ channels to open, causing a wave of polarization reversal to spread along the cell membrane | When the nerve impulse reaches the terminal branches it induces membrane proteins to allow Ca²⁺ ions to enter the cell. The Ca²⁺ ions induce vesicles containing neurotransmitter molecules to release these neurotransmitters, allowing them to bind to the next nerve cell |

Figure 8.4   Summary of the most essential features of the creation and transmission of a nerve impulse.

many different cells, so the 'decision' about whether or not the cell should fire depends on the net effect of all the different inputs, some of which will be excitatory, and some inhibitory, with the pattern of input perhaps varying all the time.

So any single nerve cell acts like an tiny automatic ballot machine, assessing the number of 'yes' and 'no' votes entering it at any one time and either firing or not firing depending on which type of vote predominates at any one time.

Here are some other complications. Many nerve cells fire more or less all the time, but undergo variations in the *rate* of their firing which depend on the number of 'yes' and 'no' votes they receive. For some neurotransmitters there is more than one type of receptor, and the different types bring about different effects within the cells they belong to. Many other membrane proteins, in addition to the major ones mentioned here, can become involved in determining, amplifying and modulating the changes in membrane depolarization upon which nervous activity depends. Some neurotransmitters work more slowly and less directly, altering the polarization of cell membranes by changing cell metabolism in ways that do not depend on the opening of membrane channels. Some dendrites can transmit nervous impulses as well as receive them, while some terminal branches can receive as well as transmit, and so on. . . .

As always, however, we can identify the crucial simplicities at the heart of the complexity of life. Nerve cells receive electrochemical signals from other cells, and each signal represents a 'yes' or a 'no' vote in an election to determine whether the cell should fire. When the 'yes' votes win the cell fires, and transmits an impulse to its terminal branches, where the impulse causes new 'yes' or 'no' signals to be sent out to the cells which these branches are in contact with. The nervous system is a massive and bewilderingly complex array of interconnected nerve cells all undergoing pulses of firing depending on the inputs they receive from one another and from other cells of the body.

The other cells of the body which send inputs into the nervous system are the 'sensory' cells of our eyes and ears and noses and tongues and skin and so on. Chemical changes in these sensory cells, in response to light, or sound, or pressure, etc., cause the cells to stimulate sensory nerve cells into firing.

The other cells of the body which can receive *outputs* from our nerve cells are the muscle cells, which are induced to undergo contraction when neurotransmitters sent out from nerve cells bind to the muscle cell membrane and initiate the changes in the chemistry of the muscle cell which get muscle contraction underway.

So the nervous system can gather up information about the outside world in the form of nerve impulses flowing from sensory nerve cells to the brain; and it can send nerve impulses out to the muscles, causing them to contract and so allow the various parts of the body to move in ways which are appropriate to the information coming in from the sense organs. What happens within the brain, however, is the most fascinating question, and it is the workings of the brain which remain the most mysterious.

The origins of mind and memory are the two central puzzles facing biologists interested in the brain. They are still at the stage of arguing amongst themselves and with philosophers about what they really mean by the 'mind' whose origin they wish to explain. They have a few plausible ideas about what might constitute a 'memory' and how it might be created, but no hard facts.

The general assumption is that mental 'sensations', 'thoughts' and 'ideas' are simply what happpens within a brain when its network of neurons undergoes specific patterns of firing. When we 'remember' something our brain is presumed to generate a pattern of firing which in some way resembles the pattern which was brought about by our initial experience of the thing we are remembering. Most theories of memory depend on molecular mechanisms which would allow the occurrence of a specific pattern of neuronal activity to make it easier for that same pattern, or a similar one, to arise again. So something which has happened once in our brains would be more likely to happen again than something which has never happened; and we can somehow encourage the repeat performances of previous neuronal activities to appear when we command them to.

That leads us into the problem of 'free will', a phenomenon which has not yet even been pinned down as a definite reality. Many people still suspect that our impression that we have some free will to control what we think and do may be an illusory trick played upon us by our mysterious minds. This is a problem I do not intend to explore, and nor shall I explore any further the unresolved problems of the origins of minds and their memories. I could easily fill a book with details of theories and opinions about these problems, but in a book of vital principles there is little place for tenuous theories and opinions. If this is an area you would like to explore, then I refer you to the books in this area listed in the suggestions for further reading on page 212.

As far as we know, our minds are created by our brains and by our brains alone; and the significant thing that happens in our brains is the transmission of nerve impulses. A nerve impulse is a physical electro-chemical wave travelling along a nerve cell; and an impulse within one cell can encourage or discourage the creation and transmission of impulses within other cells. All this 'nervous activity' is created and controlled by the activities of proteins and their specific interactions with the other chemicals of life; and, of course, the proteins are encoded by the genetic information of our genes.

In essence, that is all that our enquiring minds have been able to discover about these minds, and it may be all there is to discover.

9   The creativity of evolution

How did the glorious complexity of the living world ever come into being? As we examine the molecular mechanisms of life we uncover a baffling array of chemical systems whose integrated intricacy makes us feel we are looking at a masterpiece of purposeful design. Who, or what, was the designer?

The dogma of modern biology says that there was no purposeful designer. It says that, like all the molecular mechanisms which sustain life, the creation and continuing diversification of life proceeds automatically, powered by the blind laws of physics and chemistry. The process by which living things give rise to new and often more complex living things has become known as 'evolution', and the central working principle of evolution is known as 'natural selection'. We must examine the meaning of these two central terms before we look at their operation in the biological world.

Imagine a collection of things, each endowed with one essential ability – the ability to make other things which are very similar to themselves, but usually slightly different. Suppose that these things are surrounded by an environment which contains all the raw materials needed to allow them to make more of themselves, although the supply of these raw materials is always limited and often rather scarce. Also suppose that none of the things is immortal. Each thing, in other words, will eventually 'die' as it begins to function wrongly or falls to pieces through wear and tear. What will happen?

The existing things will make more things very similar to them-selves, although no two things will ever be precisely identical since their process of 'replication' does not produce identical copies of the existing things, but slightly different versions based on the same overall plan. Each original thing will be able to give rise to a lineage of related descendants. Old things will constantly be dying, while new ones are constantly being created. At any one time the surviving individuals within each lineage will be very similar to one another, while there will generally be greater differences between the individuals of different lineages.

Each individual thing will be at least slightly different from every other thing, although many of these differences will probably concern very minor characteristics which have little effect on the overall workings of the things. At least a few of the differences, however, will inevitably cause small but significant differences in the way in which different things operate. In a variable population, some things are bound to be slightly 'better' at making more of themselves than other things. Some might, for example, be able to perform the replication process more quickly than others. Some things will inevitably be slightly better at surviving than the others. They might, in other words, be able to 'live longer' before they eventually 'die'.

Obviously the proportion of variants in the population that are able to replicate most quickly is going to rise, while the proportion of slower replicators will fall. This will happen automatically, simply because fast replicators will obviously make more replicas of themselves, which are also likely to be fast replicators, than the slower replicators can. Similarly, the proportion of variants in the population that are able to survive longest will automatically rise, while the proportion of those that last for shorter times will automatically fall. Most 'successful' of all, of course, will be variants that can survive for a long time *and* replicate very quickly.

What will happen overall is that a population of replicating and varying things will gradually become enriched in things which can survive longer and replicate more quickly than their predecessors, and depleted of those things which can survive for only a short time and can only replicate slowly. Assuming that there are only sufficient raw materials available to sustain some finite number of things, the efficient survivors and replicators may soon drive the less efficient survivors and replicators to extinction.

Anyone watching and puzzling over these changes in the thing population through many generations would observe that the population of things is evolving. In other words, it is becoming progressively enriched in more successful forms (where success is measured simply in terms of the ability to survive and replicate). It would appear as if a process of selection was somehow driving this evolution forward, with those variants that are the best survivors and replicators being selectively preserved while the poorest survivors and replicators are being rejected. Nobody is actually doing any selecting, however. There is no laboratory technician or God constantly reaching down into the population of things to discard the poorest of the variant things and retain the best. There is no need for such a technician or God, because the variants which are the best at surviving and replicating will be automatically or *naturally* selected as the preferred progenitors of future generations, simply because their prowess at surviving and replicating automatically ensures that they and their descendants represent an increasing proportion of the thing population as time proceeds.

Natural selection is simply the preferential survival and replication of whichever things are best at surviving and replicating. Automatically, and without any magic or mystery, it allows a population of things which can make other things which are very similar to themselves, but usually slightly different, to evolve continually into successive populations of things which are ever more efficient at surviving and making other similar things.

What if the environment around the things changes? The temperature might rise, making the things more liable to disintegrate and die; some new materials might suddenly arrive from somewhere else, and so on. ... In the new environment some of the previously very successful things might find themselves at a disadvantage. Some of the previously inefficient things might suddenly find that the new environment suits them very well indeed. The requirements for success will have changed, and the population will automatically adjust in response to that change. Previously successful things which can no longer flourish under the new conditions will diminish in number and perhaps become extinct. Some previously rather unsuccessful things, which might have been on the verge of extinction themselves, might suddenly flourish; and from the continual supply of new varieties new forms of things will be naturally selected, forms which might never have had a chance to flourish in the old environment, but which can do very well indeed in the new one.

Changes in the environment will continually affect the population of things, while the activities of the things will continually affect the environment, by depleting it of particular raw materials, for example, or by enriching it in specific waste materials, and so on. A dynamic inter-relationship between the thing population and the environment will be set up, each one affecting the changes of the other.

So much for imaginary and abstract things, let us turn to the real things that matter to the evolution of life – the single cells and animals and plants which are supposed to have been created by natural selection and are supposed to be continually reshaped by natural selection.

Living things can clearly 'make other things which are very similar to themselves, but usually slightly different', which is the basic requirement for evolution by natural selection. Living things make other similar living things thanks to the ability of one DNA double-helix to undergo the process of replication which creates two daughter double-helices. This process of DNA replication allows the genome of one cell to provide the genomes needed by two cells, allowing one cell to grow and divide into two. Ultimately, it allows the genome of one human, or plant or insect or other multicellular organism to give rise to the genomes needed to allow new generations of organisms to form and develop and give rise to their own new descendants in turn.

So DNA replication provides the ultimate mechanism allowing living things to give rise to other living things, but where does the necessary variation enter the process? Remember that it is not sufficient for living things to give rise to new generations of identical living things. The new generations must include new *variations* on the basic plan, some of which make some individuals of the new generations better at surviving and replicating than their predecessors.

Somehow, the genomes of living things must be able to undergo the changes which are needed to create the variety which powers evolution. How are these changes brought about?

Any change in genetic material (usually DNA) can be described as a mutation, and if that change is passed on to future generations then these generations will consist of 'mutant' individuals, relative to the previous form. Strictly speaking, we are all mutants, because we each contain a unique genome which has never before been used to create an organism, and never will be again.

Scientists have discovered a wide range of ways in which mutations can be generated, and they are still discovering new ones. Some of the most important are summarized in figure 9.1. Remember that an organism's genome can be represented by a string of letters indicating the precise sequence in which the four bases of DNA are arranged in the DNA of that organism. Mutations bring about some alteration of that base sequence.

The simplest type of mutation involves a change to a single base. Such a change may involve one base being replaced by a different one, or one base being deleted from the base sequence of DNA, or an entirely new base being added. Simple mutations of this type can be generated by the action of chemicals or radiation upon DNA, or they can be generated when mistakes are made during the replication of DNA. Nothing in life or in chemistry is ever perfect, and every now and then a nucleotide carrying the 'wrong' base can be inserted into the new DNA molecule being made during DNA replication, or one nucleotide might be missed out, or an extra nucleotide mistakenly incorporated. A change involving only one base of a long DNA molecule might not seem a very drastic or significant mutation. Very often it will not be significant, but it can sometimes be very significant indeed.

Remember that groups of three bases form the codons which each code for the incorporation of a single amino acid into a growing protein chain. Obviously, substituting one base for another could change a codon encoding one particular amino acid into one encoding a quite different amino acid. For example, the codon TCA (which becomes UCA in mRNA) encoding the amino acid 'serine', might be changed into CCA, which encodes 'proline'. This might drastically alter the activities of the protein encoded by the gene containing the mutation, especially if the changed amino acid occupies some particularly critical site in the protein. Proline, for example, is an amino acid which tends to produce a sharp bend in a protein chain, a bend which would not have been present in the original form of the protein.

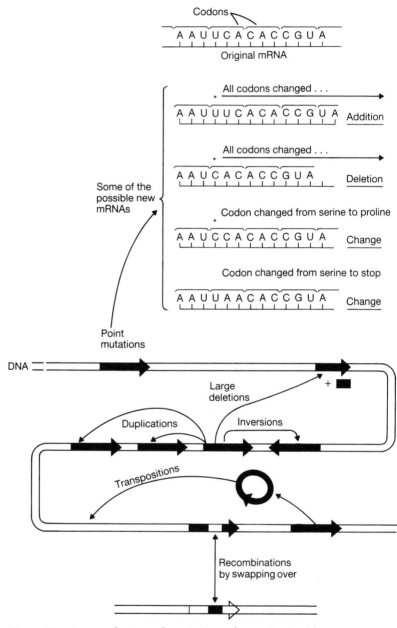

Figure 9.1   Some mechanisms of genetic change (see text for details).

More dramatically, the TCA serine codon might be changed into TAA (UAA in mRNA), which does not encode an amino acid at all, but indicates the point at which protein coding should stop. The appearance of this stop signal in the middle of a gene would at one stroke generate a new gene encoding a protein composed of only the first half of the original one. This truncated protein might fold up and act quite differently from its full-length cousin.

Mutations which cause the addition or deletion of a base can have much more drastic effects, because they can alter all the codons of the gene which come after the one affected by the mutation. As you can see from figure 9.1, the addition or deletion of a base will cause all the bases following that mutation, up to the end of the gene, to be read on the ribosome as an entirely new sequence of codons encoding a drastically different protein from the original one. Such changes are known as 'frameshift mutations', since they alter all subsequent codons by shifting the 'frame' in which the base sequence of a gene is divided up into codons.

Many of the other changes which can be inflicted upon genetic material are catalysed by various proteins, although some of them could probably occur, to a lesser extent, without any help from proteins. The fact that they are often catalysed does not necessarily mean that they are beneficial or desirable from the point of view of the cell or organism concerned. Many might be best regarded as occasional mistakes brought about by the activity of proteins that can cut and reseal genetic material. Thus large sections of DNA can become duplicated, deleted or inverted (see figure 9.1). Sometimes long sections of DNA can break loose or be copied from a cell's DNA, to wander free for a while before becoming reincorporated elsewhere, a process known as 'transposition'. Sections of DNA, especially those with broadly similar sequences, can also change places with one another in a process known as 'recombination' or 'swapping over'. This mechanism is especially adept at reshuffling genetic material between different but similar chromosomes derived from the two parents of an organism which breeds by sexual reproduction.

Almost all single cells can multiply simply by making an extra copy of their genetic material and then splitting into two cells, each retaining one copy of the genome for itself. That is how many free-living single-celled organisms reproduce, and it is also the way in which most of the cells of your own body multiply to create new tissue and replace old tissue. If you were able to look inside the cells of your own body, however, you would find that each cell apparently contains *two* sets of the human genome. Twenty-three different types of chromosome exist within human cells, but you would find that you have *two* chromosomes of each type rather than just one. It would appear, in other words, as if your cells contain two complete sets of the human 'assembly manual'. The reason for this duplication is that you are a sexually reproducing organism. You were formed when an egg cell from your mother combined with a sperm cell from your father, with each of these cells bringing with them a full set of human chromosomes – the 'maternal' set and the 'paternal' set. If we examine what happens to these chromosomes as we reproduce, we will be able to explore exactly what it means to reproduce sexually rather than by an impersonal and lonely asexual splitting in two.

Once the fertilized human egg cell has acquired its two sets of human chromosomes – the maternal set and the paternal set – this duplicated genome or 'diploid' state becomes the norm for virtually all the cells of the human body. As the egg cell divides into the trillions of cells of an adult human, each division involves the duplication of both sets of chromosomes, and the passing on of two sets to each of the two daughter cells of each division (see figure 9.2).

Something rather special, however, takes place within the cells of your body that are responsible for making your own egg cells (if you are a woman) or sperm cells (if you are a man). First, all the chromosomes in such cells become replicated, although the two new daughter chromosomes derived from each original chromosome remain joined together somewhere near their centre. Next, the matching pairs of replicated chromosomes, derived from your father and your mother, move together to form closely linked paired complexes. At this stage the maternal and paternal chromosomes are able to swap corresponding sections of DNA – a process known as recombination. This recombination yields some completely new 'hybrid' chromosomes which each contain genes from your mother and genes from your father. It is a process of genetic 'mix and match' which generates new varieties of chromosome for you to pass on to your offspring.

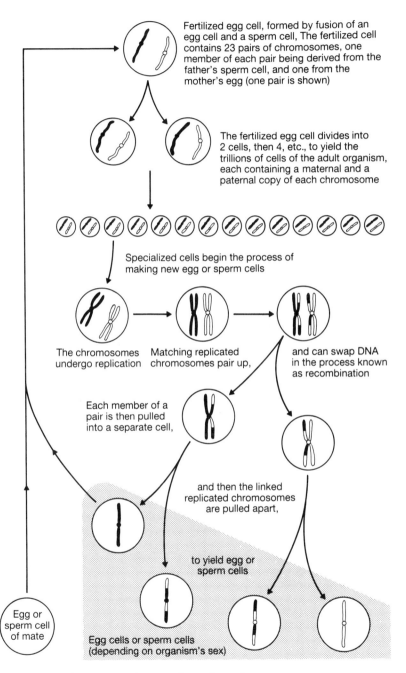

Fertilized egg cell, formed by fusion of an egg cell and a sperm cell, The fertilized cell contains 23 pairs of chromosomes, one member of each pair being derived from the father's sperm cell, and one from the mother's egg (one pair is shown)

The fertilized egg cell divides into 2 cells, then 4, etc., to yield the trillions of cells of the adult organism, each containing a maternal and a paternal copy of each chromosome

Specialized cells begin the process of making new egg or sperm cells

The chromosomes undergo replication   Matching replicated chromosomes pair up,   and can swap DNA in the process known as recombination

Each member of a pair is then pulled into a separate cell,

and then the linked replicated chromosomes are pulled apart,

to yield egg or sperm cells

Egg or sperm cell of mate

Egg cells or sperm cells (depending on organism's sex)

Figure 9.2   The genetic mechanisms of sexual reproduction.

After the recombination is complete, the cell divides in two; and as it does so contractile proteins latch on to the chromosomes and pull one member of each matching replicated pair into each newly forming cell. Whether a chromosome contains mainly maternal or mainly paternal genetic material is not distinguished in this process, so it allows large numbers of cells to produce a great variety of different daughter cells, all containing different mixtures of largely maternally and largely paternally derived chromosomes.

The two cells yielded by each such division then undergo a further division, during which the replicated chromosomes are pulled apart into single chromosomes, one going into each new cell. The cells resulting from this final division are the 'germ' cells. In other words, they are your egg cells if you are a woman, or your sperm cells if you are a man. Notice that each germ cell only contains one copy of each type of chromosome. In other words, it contains one full copy of the human assembly manual (23 chromosomes) instead of the two full copies (23 *pairs* of chromosomes) held by all the rest of your cells. This means that when you mate your egg or sperm cells will each contribute one copy of each chromosome to the fertilized egg cell which will become your child.

Sexual reproduction is clearly a very complex process of genetic mixing and recombination, but its overall effects can be stated very simply: it allows the genetic material of two different individuals to be used as the source of genetic material for individuals of the next generation. So, while asexually reproducing organisms always yield offspring which are genetically virtually identical to themselves, sexually reproducing organisms yield offspring which are genetically rather different from each of the parents involved. So sexual reproduction is a very effective means of generating genetic variation, not by inducing mutations which change the base sequence of DNA, but by shuffling different bits of the existing DNA of a species into an endless variety of new combinations.

Sexually reproducing organisms are a bit like 'genetic mixing machines'. They receive a supply of genes from each of their parents, and then they mix them up to produce new combinations to pass on to their offspring.

So there is great scope in Nature for the creation of genetic novelty:

simple 'point' mutations involving the loss or addition or change of a single base in DNA; the removal ('deletion') of large chunks of DNA; the duplication of chunks of DNA; the turning of a section of DNA back-to-front ('inversion'); the transposition of some DNA from one site on the genome to another; the recombination between chromosomes and parts of chromosomes which allows genetic material to be swapped from one piece of DNA to another; and the random mixing of genetic material achieved during the process of sexual reproduction.

The precise molecular mechanisms of all the possible changes in genetic material are rather complicated and not always well understood. The important simple point, however, is that there are lots of ways for pieces of DNA to change, grow larger, grow smaller and recombine to generate the endless genetic variation which is the raw material for evolution. The genome of an organism can be represented by a string of letters representing the base sequence of its DNA, and that sequence can change from time to time so that the genomes passed on to an organism's offspring can be slightly but significantly different from the genome that was used to create the organism itself.

So offspring are always a bit different from their parents because their genomes are slightly different, but offspring are also very much like their parents (or some blend of both parents, in the case of sexually reproducing organisms) because their genomes are copied from those of their parents, albeit imperfectly copied. Organisms whose development is directed by genomes of mutable DNA are a fine example of 'things with the ability to make other things which are very similar to themselves, but usually slightly different'.

It is a dogma of modern biology that the changes in genetic material which make offspring different from their parents are of a random nature, or at least, if they are not precisely random, they are certainly not purposefully directed towards the creation of more efficient organisms. Restrictions of chemistry may mean that mutations cannot be perfectly random, since chemistry may favour certain types of mutation reactions over others; but there is believed to be no force or mechanism at work able to encourage beneficial mutations to arise while discouraging harmful ones. Mutations, in other words, are at least as likely to produce less efficient organisms as they are to produce more efficient ones. In fact, it is widely believed that mutations are much more likely to make an organism less efficient than to make it more efficient, although many mutations, perhaps most, may be 'neutral' and have no significant effect on the working of the affected organism.

How can the continual generation of undirected and often harmful mutations possibly be the mechanism which allows life to evolve into new and ever more successful forms? The answer of the modern biologist is that the brute and blind force of natural selection can be relied upon to weed out the harmful mutations and nurture the beneficial ones. We must always remember, however, that the criterion by which natural selection judges mutations as harmful or beneficial is simply the effect of the mutations on an organism's ability to pass its genetic information on to future generations. The genetic material of organisms made less efficient by mutation (made less successful at passing on their genetic information, in other words) will face a bleak future. The organisms will either quickly die, or else their inefficiency will cause them to spawn less descendants than more efficient forms; and if these descendants inherit their parents' weaknesses then they too will spawn less descendants than more efficient forms. Over time, organisms carrying a particular harmful mutation will represent a diminishing proportion of the population, and if the mutation is sufficiently harmful they may become extinct.

Beneficial mutations, even if they occur very rarely, will automatically be 'selected for' and therefore retained in the population, since the definition of a beneficial mutation is one that helps its bearers to survive and multiply, and hence allows individuals carrying the mutation to account for an increasingly large proportion of the population.

Nature is a blind but a supremely effective engineer. Through the agency of undirected mutation she continually adjusts the structures and the mechanisms of the living things on earth. When the adjustments cause damage, the affected individuals and their lineages automatically wither away. When the adjustments create benefits, the affected individuals automatically flourish. Thus the natural selection of the beneficial effects of mutation, accompanied by the 'natural rejection' of harmful effects, allows a process of undirected change to power the continual evolution of populations of organisms into forms which are ever more able to survive and replicate in the environment around them. When the environment changes, the process may stutter as large numbers of organisms become extinct; but from the survivors, and their continual adjustment by mutation, new variants will inevitably arise that can flourish in the changed environment.

Just how effective an engineering or design process this can be is illustrated by the fact that some human designers are now realizing that a process of computer-generated evolution may be a more effective way to perfect our machines than sitting down at the drawing board and attempting purposeful design. Imagine you want to design the perfect wing for an aircraft, or a bird. You might think that the best way to do it would be to learn everything you could about aerodynamics and then draw up designs which you suspect will produce the best results. An alternative, more natural, strategy is to give a computer a very basic plan for a wing-shape, and then programme the computer to impose a series of random changes on the shape and test each one for aerodynamic improvement. The programme will reject changes which result in a worse wing, select beneficial changes, and use the wings incorporating these changes as the progenitors of new generations of wings incorporating new random changes. Artificial 'breeding' processes like this have been used to generate highly efficient wing-shapes far more quickly than any human designer can.

Many people, on becoming aware of the intricate complexity of life, declare that 'there must have been a designer to think all this up and put it together'. The lessons from the new field of 'evolutionary engineering' suggest that no designer was needed. Indeed, they suggest that a process of undirected mutation may be the most efficient and innovative designer of all, provided some mechanism is available to reject harmful changes in design, select beneficial ones, and incorporate the beneficial ones into the progenitors of new varieties.

Natural selection is believed to be the mechanism which allows blind physics and chemistry to 'design' the intricate complexity of life. The basic instructions for making living things are encoded in a sequence of chemical bases which can be represented by the letters A, T, G and C. Modifications in the design of life consist of alterations and additions to and sometimes deletions from the base sequences at the heart of all life. Changes to the instructions result in changes in the battery of proteins which the instructions cause to be created; and changes in the battery of proteins produced by living things cause the changes in the structure and functions of living things which allow evolution to proceed.

The ascent of life

Nobody knows exactly how or where life began. It is an article of faith amongst modern biologists that life, or at least our kind of life, originated on earth as a result of the blind processes of physics and chemistry. A few dissident scientists believe that this is an unnecessarily parochial view, and that life may well have originated elsewhere in space before the earth was born. 'Seeds of life', in the form of simple micro-organisms or their spores, could then have colonized the earth soon after its origin. Those who believe in such a process, however (or at least most of them), still believe that life must originally have begun thanks to the actions of blind physics and chemistry in a suitable location somewhere in the universe.

We should not be too dogmatic about things we can never know for certain. Some people think life was created by some superior being, or force or principle known by such names as 'God' or 'Allah'. Such beliefs, of course, merely replace the problem of the origin of life with the problem of the origin of God. Some people think that there is infinitely more richness and complexity to the physical universe than we are aware of. They suggest there might be unseen dimensions and other hidden spaces and processes which are as invisible to us as the earth and the sky and the stars are to a micro-organism inhabiting some dark hydrothermal vent on the floor of the deepest ocean. Without knowledge of these secrets, they contend, we can never fully explain either our origin or the true essence of our being.

Elementary textbooks of biology relate a simplistic tale about the origin of life on earth which may be true, or partly true, but may be quite false. They describe a prebiotic world in which a rich mix of simple chemicals was formed and underwent reactions which pro-duced the chemical building blocks of life today. They tell of short chains of nucleic acids forming spontaneously, and then becoming able to encourage their own replication in a manner similar to the replication of DNA today. If such nucleic acids did form spontaneously on the early earth, and were able to encourage their own replication, then evolution guided by natural selection would have begun. These primeval nucleic acids would have been the first things able to make other things which are very similar to themselves, but usually slightly different. The first letters in the code of life would have become linked together. Undirected mutations in the first self-replicating nucleic acids would have allowed new, more efficient, nucleic acids to evolve. Some of these might have been 'more efficient' because they encouraged other components of the primordial world to cluster around the nucleic acids and form the earliest simple 'cells'. At some point, of course, some of the nucleic acids would have become capable of the great 'trick' of encouraging specific protein molecules to form. Nucleic acids would have begun to encode proteins, and, with the awesome catalytic powers of the proteins available, life would have really been on its way.

It is an appealingly simple tale. Many think it is far too simple to be taken seriously. Some suggest alternative scenarios in which modern life evolved from completely different beginnings, in which the nucleic acids such as DNA and RNA played no immediate part. Some suggest that the earliest living (or at least evolving) things were not composed of organic chemicals, but were composed of inorganic minerals which eventually gave rise to the organic chemicals which displaced them.

This is not the place to enter into a deep analysis of the mystery of the origin of life. I immodestly refer readers interested in this fascinating puzzle to my book *The Creation of Life* (Basil Blackwell, 1986) which is devoted entirely to this problem. For our present purposes it should be sufficient for me to say that I am not able to reveal to you how life began, because I do not know for sure, and neither does anyone else. Scientists have developed a few plausible ideas on the subject, and quite a few less plausible ones. The attempts to recreate the chemistry involved in life's origin are at a very early stage and have met with no really dramatic and convincing successes. It will be some time, at least, before we can describe the precise chemistry of life's origin with the same confidence as we can describe the chemistry which sustains life today. That is no disgrace. It is difficult to describe with precision events which occurred at least 4,000 million years ago when nobody was around to witness them.

The evolution of the first life-forms into the complex and multifarious creatures of today is a long saga spanning billions of years but with many highlights. We can relate, with increasing confidence, the highlights of the tale from the time of the earliest single cells onwards, because fossilized remains of many of the actors in the drama are available for us to study. I have outlined the basic mechanism which drove (and continues to drive) the development of the plot: evolution powered by the natural selection of undirected mutations in the base sequence of DNA, all made possible by the physics and chemistry with which our universe is endowed. Yet again I must emphasize that this is a book of principles rather than of details. So, having outlined the basic principles involved in both life and the evolution of that life, let me scan through the story of evolution with the 'fast forward' button firmly pressed, pausing only occasionally to give you a glimpse of certain key highlights.

The first cells presumably consisted of little more than short and simple nucleic acids, able to encourage the manufacture of a few simple and not particularly efficient proteins, all surrounded by a simple lipid bilayer membrane. The raw materials needed to sustain these cells would be taken up from the surrounding environment and subjected to very little processing, since the enzymes of complex metabolism were yet to evolve. All the reactions involved in the life of the cells would need to be energetically favourable in their own right, or would need to be coupled to other energetically favourable reactions in very simple ways. The cells would largely be sustained by reactions between high energy chemicals gathered up from the environment, and which were in high energy states because they had been put there by reactions driven by the energy the earth receives from the sun, or by other sources of energy such as volcanic activity, radioactive decay, or the violent impact of meteorites with the earth's surface.

As evolution proceeded, the nucleic acids at the heart of cells would have grown longer and more complex, and able to give rise to many more proteins. These proteins would have evolved into ever more efficient catalysts able to process the raw materials taken up from the environment in ever more complex and useful ways. ATP, or some similar high energy molecules, would have become adopted as the 'energy currency' of cells, being manufactured at the expense of reactions in which high energy materials from the environment fell down into lower energy states, and used (by the principle of chemical coupling) to power energy-requiring reactions useful to the cell.

Eventually, some cells must have arisen whose proteins and membranes and metabolites allowed them to capture the energy of sunlight directly and use it to manufacture ATP, and hence to drive forward all the energy-requiring reactions of the cell. Photosynthesis, in other words, would have begun.

One of the most significant developments in the evolution of life on earth must have been when some of the early simple cells began to live within other cells. Some of these gatecrashers within other cells would then have slowly evolved into the mitochondria and chloroplasts of present-day life – descendants of cells which took up residence within other cells and then became so dependent on their hosts, and also so useful to them, that neither hosts nor gatecrashers could afford to live apart.

Another momentous phase began when cells started to interact with one another before giving rise to progeny, rather than each cell simply splitting in two. When cells came together before reproducing they became able to create offspring containing genetic information derived from both parents – the process of sexual reproduction began.

A little while after the origin of sex, and perhaps thanks to the increased possibilities for variation and evolution which sex made possible, cells began to live constantly attached to other cells in the form of the first simple multicellular organisms. Cells, which had previously been aggressively independent individualists, discovered the advantages of communal life.

The fossil record reveals how rapidly multicellular life prospered and diversified. There began a process of increasing 'division of labour' between the different cells of multicellular life – some becoming specialized for gathering up food for all the other cells, some becoming specialized at determining the shape and structure of the organism, some becoming specialized for defence, and so on.

Multicellular creatures began to develop distinct tissues and organs in which groups of related types of cells performed specific specialized tasks for the benefit of the organism as a whole.

They began to develop a distinct 'body cavity' or 'coelom' in which many of the organs were to be found. They began to develop hard shells and internal skeletons, and nerves and powerful muscles which allowed them to swim through the water and crawl on to the land. Locomotion and nerves became the virtual monopoly of the animals, while the plants were also developing in their own ways and also colonized the land. Although the animals were able to rush about and eventually begin to think about what they were doing, the plants kept their monopoly on the tricks of photosynthesis, which meant that animals had to eat them, but which also ensured that the animals could never survive without them.

It has taken several billions of years, at least, for evolution driven by the natural selection of undirected novelties in the base sequence of life's nucleic acids to create the creatures of the modern world; and the process of evolution will continue to craft new forms of life to cope with the new challenges of the future.

Throughout the long course of evolution all the creatures of the earth have been sustained by the same simple vital principles: the forces and laws of physics which make chemical reactions proceed, which allow nucleic acids to direct the manufacture of the proteins, which bring order and specificity to the chemistry which makes life work.

Outside it all, 93 million miles away, the giant furnace of the sun continually throws out the energy which allows the operation of these principles to raise up the dust of the earth into the energetic dance of life.

10   Principles to live by

We are living through privileged times indeed. Times in which some answers can at last be given to questions which have troubled every human since human life began: What am I? Why am I here? How do I work?

Of course, the answers available at the moment are not completely satisfactory, and perhaps they never will be. When creatures as proud as ourselves look inwards to discover what we really are, we may always be somewhat disappointed at the answers we obtain. The most common reaction of anyone first instructed in what science has discovered about the human body is to say 'but surely there must be more to me than that!' At the moment, all that science can truthfully say in reply is that there may well be more that remains hidden, but there may well not be.

Regardless of what secrets, if any, remain, the endeavours of countless humans over hundreds of years have provided us with a deep and fascinating insight into what life is and how it works. As a summary, and a reminder, here is a concise list of the most vital principles:

All the structure of the universe, and all events within the universe, including all the activities of living things, depend on five basic phenomena - space-time, matter, charge, force and energy.

Space-time is the four-dimensional world in which all other phenomena are found. It consists of the familiar three dimensions of space united with the one dimension of time, to create a place in which every thing and every event occupies its own unique location.

Matter is the stuff from which everything with mass (i.e. every material object) is made.

Charge is a mysterious phenomenon which we can describe but not explain. The most familiar type of charge is electrical charge, which comes in two forms – positive and negative. A force of attraction exists between any two objects carrying charges of opposite signs, while a force of repulsion is experienced between objects carrying charges of the same sign.

A force is something capable of altering the motion of an object. In other words, it is a phenomenon which can push or pull at an object.

Energy is some sort of force resistance or antiforce, able to resist the pushes and pulls of the fundamental forces. It is often formally defined as the ability to do work, with work being any process which involves some resistance against a fundamental force.

Matter is composed of atoms, which are themselves composed of protons, neutrons and electrons. The protons are positively charged, the neutrons are electrically neutral and the electrons are negatively charged.

Atoms contain equal numbers of protons and electrons, making them electrically neutral overall.

An atom's protons and neutrons are clustered together into a central nucleus, surrounded by the electrons in their orbitals.

Each electron has the required amount of energy to keep it 'orbiting' around the atomic nucleus, despite the force of attraction between the electron and the nucleus.

No more than four fundamental forces do all the pushing and pulling and changing required to make the universe work. These four are: the force of gravity, which attracts all matter towards other matter; the force of electromagnetism, responsible for the attraction between opposite electrical charges, the repulsion between like electrical charges, and the phenomenon of magnetism; the strong nuclear force, responsible for binding protons together in the atomic nucleus; and the weak nuclear force (probably another aspect of the electromagnetic force) responsible for some forms of radioactive decay within atoms.

Atoms can undergo chemical reactions to form molecules (particles composed of two or more atoms chemically bonded together) and ions (atoms or molecules which have become electrically charged by losing or gaining one or more electrons).

All chemical reactions involve interactions between atoms and/or molecules and/or ions.

Molecules consist of atoms joined together by covalent bonds, formed when electrons become shared between the bonded atoms. Equal sharing produces a pure covalent bond. Unequal sharing produces a polar covalent bond, in which one of the atoms carries a slight negative charge (δ^-) while the other carries a slight positive charge (δ^+).

Ions can be bonded together in ionic compounds by ionic bonds. An ionic bond is essentially just the force of attraction between positively and negatively charged ions.

In all chemical reactions atoms and/or molecules and/or ions meet up and interact with one another, causing chemical bonds to be broken and/or made to form new chemicals composed of different combinations of atoms and/or molecules and/or ions.

Chemical reactions are driven forward by the tendency of energy to disperse towards an even distribution (the second law of thermodynamics). They are the result of chemicals interacting and rearranging their electrons in ways that move the chemicals towards energy states that are compatible with the energy of their surroundings.

Chemicals come into contact with one another because they are all constantly moving with the energy we call heat.

This thermal motion turns much of the chemical microworld into a molecular mixing bowl.

The combination of random thermal motion and simple specific interactions between chemicals when they meet can generate complex and seemingly purposeful change.

Thermal motion can break chemical complexes apart, as well as bring their components together.

The basic functional unit of all living things is the cell, a structure in which a lipid bilayer membrane encloses all the chemicals needed to allow the cell to live and grow and multiply.

Multicellular organisms, such as humans, begin life as a single cell. That cell then undergoes a series of divisions which lead to the creation of all the cells of the adult organism.

The structures and activities of all cells are believed to be entirely the result of chemical reactions between the chemicals they are made up of.

The chemical components of living cells can be divided into five main categories – DNA, RNA, proteins, membranes and metabolites.

Proteins are large molecules composed of different amino acids linked together. The proteins of a cell are the molecular workers which do most of the tasks involved in constructing and maintaining cells.

Proteins can act as enzymes; can play structural roles; can form contractile assemblies; can act as transporters of various chemicals; can act as chemical messengers; can act as the receptors of chemical messages; can act as chemical gates and pumps in the cell membrane; can act as chemical controllers – controlling the activities of other proteins, and of DNA and RNA; can act as defensive weapons; and can also play other lesser assorted roles in the life of a cell and an organism.

Proteins achieve their many effects by selective binding to other chemicals, followed by acts of catalysis and/or conformational change.

In order to perform their acts of selective binding, catalysis and conformational change, proteins must fold up into a complex three-dimensional structure which is determined entirely by the amino acid sequence of each newly formed protein chain.

This protein folding proceeds spontaneously, governed by forces of electromagnetic attraction and repulsion between parts of the protein and parts of the water molecules surrounding the protein.

When a protein adopts, automatically, its final folded form, it presents a very specific chemical surface to the environment around it. The precise shape and chemical nature of this surface determines what the protein will be able to do.

Some proteins must become bound to chemical cofactors, or must undergo some permanent chemical modification, or must aggregate into multi-subunit complexes, before they can perform their tasks within the cell.

DNA is the chemical, stored in the cell nucleus, which contains the genetic information that determines what proteins a cell will contain.

The DNA in the nucleus is in the form of a double-helix, in which two individual DNA strands are wound around one another.

The two strands of a double helix are held together by weak bonds between complementary bases of the DNA.

Four bases (represented as A, T, G and C) are found in DNA. The As and Ts of different strands can pair up to hold a double-helix together, as can the Gs and Cs.

So the double-helix is held together by the base pairs A-T and G-C.

The genetic information of life is encoded in the base sequence of DNA.

The first step in the decoding of this genetic information is its copying (transcription) into single-stranded messenger RNA which is complementary to one strand of the corresponding double-helix.

The messenger RNA binds to ribosomes in the cytoplasm, where its specific base sequence directs the creation of a protein with a specific amino acid sequence.

The link between base sequences in DNA and RNA, and the amino acid sequences of the proteins these nucleic acids encode, is that groups of three bases (codons) on DNA or RNA correspond to each individual amino acid of a protein.

As a ribosome moves along a messenger RNA it exposes successive codons at a critical site on the ribosome, allowing transfer RNAs to bind to the codons via their complementary anticodons. Each transfer RNA brings with it the amino acid encoded by the codon to which the transfer RNA can bind. As successive amino acids arrive at the ribosome, they are linked up into the growing protein chain.

Thus a specific sequence of bases in a DNA double-helix can, via an RNA intermediate, direct the manufacture of a protein composed of a specific sequence of linked amino acids.

Each section of a double-helix which encodes a specific protein molecule can be called a gene.

Genes are found at intervals along massive lengths of double-helical DNA known as chromosomes.

DNA double-helices can undergo replication when their individual strands separate, allowing new complementary strands to be assembled on each original strand as directed by the rules of base-pairing. This is the molecular mechanism which underlies heredity, allowing new copies of an organism's genetic information to be made and passed on to its offspring.

Nucleic acids (DNA and RNA) are able to direct the manufacture of specific proteins, and to undergo the replication which makes them the agents of heredity, thanks to their ability to act as genetic moulds upon which new complementary strands of nucleic acid can be assembled.

Life results from the mutually interdependent interaction between proteins and nucleic acids. Proteins do most of the chemical work, while nucleic acids store the genetic information needed to specify the structure and activities of future generations.

The central dogma of our molecular explanation of life is that genes (of DNA) make RNA, which makes protein.

But the phenomenon of life is really a self-sustaining global automaton in which DNAs and RNAs and proteins all depend on one another's activities. Proteins act on DNA to make new DNA and RNA, and then on RNA to make new proteins, which can in turn take the place of the old proteins to act on DNA to make new DNA and RNA, then more proteins, and so on. . . .

The membranes of living cells are all based on the lipid bilayer structure. They form a boundary to all cells, and they cordon off specific regions within a cell into distinct organelles.

The metabolites of a cell are all the chemicals used by the cell, and made by the cell, in order to allow it to live. Metabolites are taken up into the cell, broken down, processed, stored, exploited or manufactured due to the ability of proteins to interact specifically with the metabolites and catalyse chemical reactions in which they participate.

A complex system of membranes and proteins and other chemicals in the chloroplasts of plant cells captures the sun's energy and uses it to generate high energy compounds, especially ATP.

Animals can capture the energy of the sun indirectly, by eating plants (or by eating other animals which are themselves ultimately sustained by plants) and allowing the high energy chemicals they eat to react with oxygen to reach low energy states, a process which drives forward the manufacture of ATP.

ATP serves as the main energy currency of cells. Energy-requiring processes within the cell can be made to proceed by coupling them to the energy-releasing break-up of ATP, so that a little energy is released into the environment overall.

ATP does not itself power all the energy-requiring reactions of cells directly; but all energy-requiring reactions are made to proceed by being coupled to some other reaction which releases the energy required.

Many of the chemicals of living things are able to bind to and alter the activity of some of the other chemicals of living things. Such interactions allow the activities of the cell (and the organism) to be integrated, adjusted and controlled to match changing needs and changes in the internal and external environment.

The populations of living things of today have evolved from the populations of living things of the past, and will evolve into the life of the future, thanks to the natural selection of undirected novelties generated in the base sequence of organisms' DNA.

And all the principles listed so far are encompassed within the three central secrets of life:

The electromagnetic force interacts with the energy of the world to make chemical reactions proceed.

Chemicals called nucleic acids are able to form and to direct their own reproduction and also to direct the production of chemicals called proteins.

Proteins, directly or indirectly, promote the specific chemical reactions needed to create and sustain all life.

11   Mystery remains

We all have a tendency to overestimate our own achievements and forgive and forget our shortcomings. This applies to scientists and science as a whole, as much as to anyone else. The knowledge which humanity has accumulated about the mechanisms of life is both impressive and convincing, but there are some large and important gaps. Our admiration of what we have been able to find out should not blind us to the many mysteries which remain.

Sadly, the deepest mysteries concern the questions we would most like to be able to answer.

We are all living minds, and yet science cannot tell us exactly what a mind is, or what a thought, an idea or a sensation in our brain is. It can offer no really satisfactory answer to the puzzle of how inanimate matter composed of atoms and molecules and ions can give rise to the consciousness which is able to identify and confront such puzzles.

Neuroscientists might protest and claim that they know an enormous amount about the brain; and so they do, but they do not really have answers to the questions most people would like answered. My survey of the principles of the operation of the nervous system was cursory in the extreme. A neuroscientist could easily fill a whole book with further details about what is known to happen within nerve cells and brains, and all written at the same basic level as this book; but I suspect it would still leave you feeling disappointed, and convinced that no satisfactory explanation of our minds and their memories has yet been found.

To expect a satisfactory explanation of the nature of our own conscious minds may be asking too much. It seems quite likely that our minds would be unable to comprehend a full explanation of themselves, since the whole may never be able to analyse and understand anything more than a part, or parts, of itself. We may simply have to accept that consciousness and thought and memory, and the mysterious power of free will, is what happens when a collection of nerve cells undergoes specific patterns of nervous activity. We may never really understand why.

How one feels about this field of enquiry is a very personal thing, affected by one's religious beliefs, by one's level of expectation and credulity, and by such intangible things as the level of mystery and awe, or nonchalant acceptance, one experiences when faced with the fact of one's own existence.

I referred to the mind earlier as 'the hidden inner chamber of biology'. I personally feel that it is a hidden chamber which we will be able to identify and examine much more fully in the not too distant future. There are surely some crucial unknown, but knowable, principles of neurology which will eventually need to be added to the list of the previous chapter. I doubt, however, if we will ever arrive at any explanations which will make our conscious minds feel fully satisfied when they ask themselves 'what am I?'

Whatever the truth about the origin of the mind and its memories and apparent free will, there are certainly many unsolved mysteries left to challenge the ingenuity of those researchers who choose to investigate the molecular mechanisms of the brain and the mind.

There are plenty of mysteries in other areas as well. The complex process of development, in which a fertilized egg cell gives rise to the massive and exquisite chemical sculpture we call a human, remains an area of great mystery. At the moment we can say that, *in principle*, this development must involve a series of interactions between genes and the chemicals of their cells and the environment which result in the right genes being active at the right times in order to produce a healthy human rather than a lifeless chemical mess. Almost all the details, however, remain to be uncovered; and until they are uncovered it remains possible that there are mechanisms involved in the development of multicellular organisms which we have no inkling of at the moment.

If I were asked to name the major mysteries facing biology today, I would say: the mind and its memories, and the control of the development of multicellular organisms from fertilized eggs or seeds.

There are plenty of others though. For example, we do not really know in any significant detail why organisms 'grow old' and die. We know lots of things about what happens to us as we age and die, but little about the fundamental molecular mechanisms behind it. A related problem is the question of why we develop degenerative diseases, such as cancer, heart failure, strokes and so on. Again, we know a great deal about what happens as such diseases develop, but not so much about why they develop.

There are also many mysteries concerning the precise mechanisms of many of the vital processes examined in this book. In just about every case, we know what happens overall, and have a good understanding of how and why the major steps proceed, but a great many details and subtleties remain to be uncovered.

So plenty of mystery remains within the organs, tissues and cells of living things to keep us intrigued and occupied for a long time to come. The fact that books such as this one are able to catalogue the most essential features of the physics, chemistry and biology of life does not mean that humanity has learned all there is to know about life on earth, or has even got close to learning all there is to know. It simply means that, after thousands of years of ignorance, the past hundred years or so have seen us make a good start in our attempts to understand what we are and why we are here.

We have made a good start and have identified some, but almost certainly not all, of life's vital principles; but there is a lot of work yet to be done and several major surprises probably lie in store.

Glossary

This is a summary of the main technical terms used in this book *in the context in which they are used in the book*. It should not be regarded as providing rigorous 'dictionary definitions' of all the terms, since some of the terms can be used in various other ways and contexts. This glossary is offered merely as a simple aid to readers as they work through the book.

Actin A globular protein, many molecules of which aggregate together to form the actin filaments involved in muscle contraction.

Adenine One of the bases found in DNA and RNA. It pairs with thymine in DNA and uracil in RNA to form the A-T or A-U base-pairs.

Adenosine diphosphate (ADP) The chemical which combines with phosphate to make ATP, and which is formed when ATP breaks up into ADP and phosphate.

Adenosine triphosphate (ATP) A high energy chemical which serves as the main 'energy currency' of the cell. Its energy-releasing conversion into ADP and phosphate supplies the energy needed to make many energy-requiring reactions of the cell proceed.

Amino acids The simple chemical building blocks of all proteins. Twenty different amino acids are available to make proteins, in which they are linked up into long chains of specific 'amino acid sequence'.

Anabolism A general term covering all the processes within cells which involve the manufacture of new and usually complex molecules from simpler ones.

Antibodies Defensive protein molecules produced by many complex multicellular organisms. Antibodies can bind very specifically to invading micro-organisms and foreign substances of many kinds, and so bring about their neutralization and elimination.

Anticodon A group of three bases on a transfer RNA molecule which can form base-pairs with a complementary codon on messenger RNA, and so allow each codon to specify the incorporation of a particular amino acid (carried by the transfer RNA) into a growing protein chain.

Antigens The specific chemical groupings to which antibodies can bind.

Asexual reproduction Form of reproduction which does not involve the sexual union of the germ cells of two different organisms.

Atoms The basic particles of chemistry, composed of protons, neutrons and electrons.

Axon The long process of a nerve cell which conducts nerve impulses away from the cell body and towards the cell's terminal branches.

Bacteria Simple micro-organisms which lack a nucleus and all the other organelles found in plant and animal cells.

Base (of DNA or RNA) Chemical components of the nucleotides which are linked together to form nucleic acids. Each base can form a base-pair with a specific complementary base on another strand of nucleic acid.

Bond A chemical linkage between two atoms or ions.

Carbohydrates Common chemicals of life, composed of carbon, hydrogen and oxygen, although sometimes modified by the addition of other types of atom.

Carbon fixation cycle The metabolic cycle by which the cells of photosynthetic organisms can incorporate the carbon atoms of carbon dioxide gas into the complex organic chemicals of life.

Catabolism A general term covering all the processes within cells which involve complex chemicals being broken down into simpler ones.

Catalyst A substance which speeds up a chemical reaction while itself remaining unchanged, overall, in the process.

Cell The basic unit of life. A cell consists of a membrane-bound sac of watery fluid, containing all the chemicals which allow the cell to live and reproduce.

Cell membrane The fatty lipid bilayer membrane which forms the boundary of all cells.

Charge A mysterious phenomenon which makes objects carrying the charge feel the effects of some fundamental force. For example, objects carrying electric charge (positive or negative) feel the effects of the electromagnetic force.

Chlorophyll The green pigment of plant cells (and other photosynthetic organisms) which captures the energy of light during photosynthesis. (There are actually several different types of chlorophylls.)

Chloroplast The organelle of plant cells in which photosynthesis occurs.

Chromosome An object in the nucleus of a cell, composed of a portion of the cell's DNA plus various proteins bound to the DNA. The entire genome of a cell is distributed between its various chromosomes.

Codon A group of three bases which codes for the incorporation of a specific amino acid into a growing protein chain, by forming base-pairs with a complementary anticodon of a transfer RNA.

Coenzyme An organic compound which becomes bound to a specific enzyme and helps the enzyme to achieve its act of chemical catalysis.

Cofactor Any molecule or ion which becomes bound to a specific protein and helps the protein to perform its biological function.

Compound Any chemical which is composed of two or more types of atoms, or ions, bonded together by covalent, polar covalent or ionic bonds.

Covalent bond A chemical bond between two atoms which is formed when electrons become shared between the atoms involved. Equal sharing results in a pure covalent bond, unequal sharing in a polar covalent bond.

Cytoplasm The insides of a cell, apart from its nucleus.

Cytosine One of the bases found in DNA and RNA. It pairs with guanine to form the G–C base-pair.

Cytoskeleton A complex network of molecular fibres and filaments which permeates the cytoplasm of most cells.

Cytosol The fluid inside cells, excluding all the organelles and the fluids these organelles contain.

Dendrite A thin extension from the body of a nerve cell, whose major role is to receive nervous inputs from other nerve cells.

Differentiation The process whereby cells in the embryo of a multi-cellular organism become specialized (differentiated) to perform specific roles as specific types of cell in the adult organism.

Diploid A term used to describe the genomes of cells and organisms which contain *two* versions of each type of chromosome.

DNA Deoxyribonucleic acid – the nucleic acid which carries the genetic information of most forms of life.

Double-helix A structure formed when two complementary DNA molecules become wound around one another in the form of two intertwined helices or spirals. This is the structure of the DNA which forms the genomes of all cells.

Egg cell The germ cell of a female sexually reproducing organism.

Electric charge The positive or negative charge which makes objects carrying the charge experience the effects of the electromagnetic force.

Electromagnetic force One of the fundamental forces of nature. It is responsible for the force of attraction between objects carrying electric charges of opposite signs, and the force of repulsion between objects carrying electric charges of the same sign; and it is also responsible for the phenomenon of magnetism.

Electromagnetic radiation A form of energy, including visible light, radio waves, infra-red rays, and x-rays, which can be transmitted through space to influence the electromagnetic behaviour of objects it interacts with. The energy we receive from the sun consists of electromagnetic radiation.

Electron A tiny sub-atomic particle which carries negative electric charge. Found 'orbiting' around the nucleus of an atom.

Electron transport chain A series of proteins and other chemicals which can transfer electrons between them. Found, for example, in the thylakoid membrane of chloroplasts, where an electron transport chain is intimately involved in capturing the sun's energy and using it to make ATP.

Endocytosis A process by which cells can take up materials from the environment by capturing them in a pit formed in the cell membrane, which then closes up and breaks off into the cell as a membrane-bound vesicle containing the material from the outside.

Endoplasmic reticulum A network of membranes within the cell, responsible, amongst other things, for cordoning off some newly formed proteins from others and allowing certain proteins to be dispatched to particular locations within or outside of the cell.

Endosome An internal organelle of the cell, involved in the processing of some of the raw materials taken into the cell.

Energy An abstract idea which corresponds to some sort of force resistance or antiforce, able to resist the pushes and pulls of the fundamental forces. Energy is often formally defined as the ability to do work, with work being any process which involves some resistance against a fundamental force.

Enzyme A protein molecule which acts as a biological catalyst, catalysing a specific chemical reaction involved in the chemistry of life.

Evolution The process by which the earliest living things are believed to have given rise to all later forms of living things, and by which current life will generate the life-forms of the future. Evolution is believed to be caused by the natural selection of beneficial novelties generated in organisms' genomes by random, or at least undirected, mutation.

Expression of genes The complete decoding of the genetic information of a gene into a functional protein molecule, involving both the transcription (into RNA) and then the translation (into protein) of that genetic information.

Feedback inhibition A process whereby the product of some step in a metabolic pathway is able to inhibit an earlier step of the pathway.

Fermentation The extraction of energy from high energy compounds by living things, using metabolic processes which *do not* involve combining the high energy compounds with oxygen.

Fertilized egg cell The cell produced by the union between the egg and sperm cells of sexually reproducing organisms. Once formed, the fertilized egg cell begins the process of cell division which yields all the cells of the adult organism.

Fundamental forces The four basic forces which seem to do all the pushing and pulling and changing required to make every phenomenon in the universe take place.

Gene A region of DNA which encodes one protein molecule, or one functional RNA.

Genetic material The nucleic acids DNA or RNA, which store genetic information in the form of specific sequences of the bases which are strung out along their length.

Genome All the genetic information of an organism, i.e. all its DNA.

Germ cells The egg or sperm cells of sexually reproducing organisms, which unite to form the fertilized egg cell which then begins to divide to produce all the cells of the adult organism.

Glycoproteins Protein molecules which have been modified by having carbohydrate groups attached to them.

Golgi apparatus Membrane-bound vesicles which break off from the endoplasmic reticulum, allowing them to carry the protein molecules they contain to various specific locations.

Gravity One of the fundamental forces of nature. It is responsible for the force of attraction between all material objects.

Guanine One of the bases found in DNA and RNA. It pairs with cytosine to form the G-C base-pair.

Heat A measure of the kinetic energy of motion possessed by the particles of a substance. Loosely, a measure of how fast the particles of a substance are moving.

Heme A cofactor which becomes bound to the protein of hemoglobin and serves to bind to the oxygen molecules which the hemoglobin transports around the body.

Hemoglobin The protein (including its heme cofactor) which transports oxygen around the body.

Hormones Chemicals which are released from certain cells in the body, in order to travel around the body and bind to and influence the activity of other cells.

Hydrogen bond A weak chemical bond resulting from the electromagnetic attraction between a hydrogen atom carrying a slight positive charge (by virtue of its being at one end of a polar covalent bond) and another atom carrying a slight negative charge (by virtue of its being at one end of another polar covalent bond). Hydrogen bonds are the bonds which hold together the base-pairs of complementary nucleic acids. They are also crucially involved in the folding of protein molecules.

Hydrophilic Water-loving. Used to describe substances which readily mix with water.

Hydrophobic Water-hating. Used to describe substances which do not readily mix with water.

Introns Sections of the RNA copy of a gene which do not code for protein, but which are found in between sections which do code for protein. The introns must be spliced out of precursor RNAs during the generation of the mature messenger RNAs.

Ions Electrically charged particles formed when atoms or molecules lose or gain electrons. Ions can be positively charged, due to a loss of electrons, or negatively charged, due to a gain of electrons.

Kinetic energy The energy of motion associated with all moving objects.

Lipid A general term for a wide variety of fatty substances found within cells.

Lipid bilayers Structures formed when two layers of lipids line up back-to-back. The basic structure of lipid bilayer membranes.

Lysosome An internal organelle of the cell, involved in the processing of some of the raw materials taken into the cell, and the degradation of cellular wastes.

Matter The stuff from which everything with mass (i.e. every material object) is made.

Membrane potential The voltage difference across a cell membrane.

Messenger RNA (mRNA) The RNA copy of a gene which becomes bound to a ribosome and directs the manufacture of a specific protein.

Metabolic cycle A series of enzyme-catalysed reactions in which some chemicals are repeatedly generated in a cyclical manner.

Metabolic pathway A series of enzyme-catalysed reactions which combine to convert some initial starting material into some final product.

Metabolic turnover The process of continual degradation and renewal of all the components of living cells.

Metabolism The sum total of the chemical activities occurring within a cell or organism.

Metabolite A chemical which participates in metabolism, usually as a substrate, product or cofactor of an enzyme-catalysed reaction.

Mitochondrion The organelle within cells in which chemicals derived from an organism's food (or high energy stores) are combined, indirectly, with oxygen. The energy released in the process is used to manufacture ATP.

Molecule Any chemical particle composed of more than one atom, whose individual atoms are all bonded together by covalent or polar covalent bonds.

Mutation In its most general usage, a mutation is any change imposed upon the DNA of an organism's genome.

Myosin A multi-subunit protein which forms the myosin filaments involved in muscle contraction.

NADP⁺ Nicotinamide adenine dinucleotide phosphate (oxidized form). (See NADPH, below.)

NADPH Nicotinamide adenine dinucleotide phosphate (reduced form). An important coenzyme. It is formed from NADP⁺ during photosynthesis, along with ATP.

Natural selection The natural survival and proliferation of genes and organisms carrying mutations which help the affected genes or organisms to survive and spread. Believed to be the fundamental principle responsible for directing the course of evolution.

Nerve impulse A pulse of electrochemical change which spreads across the membrane of a nerve cell.

Neurotransmitter A chemical released from a nerve cell and which can then bind to neighbouring nerve cells and either encourage or discourage them to fire, and perhaps also modulate their activity in other ways.

Neutron A sub-atomic particle which carries no overall electric charge. Found in the nucleus of an atom.

Nucleic acids Chemicals which form the genetic materials of life – DNA and RNA. All nucleic acids are composed of many nucleotides linked into a chain.

Nucleoside triphosphates The immediate raw materials of the manufacture of new nucleic acids. They all consist of a nucleotide combined with two phosphate groups.

Nucleotides The molecules which form the individual links in all nucleic acid chains. Each nucleotide consists of a base, a sugar and a phosphate group.

Nucleus (atomic) The tiny central body of an atom, where all the atom's protons and neutrons are clustered together.

Nucleus (cell) The roughly central organelle of a cell which contains the cell's genome of double-helical DNA.

Orbitals Regions of space within atoms in which electrons can be found (with a maximum of two electrons per individual orbital).

Organ A multicellular part of an animal or plant which forms a distinct unit specialized to perform some particular function, e.g. the heart, the liver, the brain, etc.

Organelle Any membrane-bound vesicle within a cell, containing its own particular collection of chemicals cordoned off from the rest of the cell.

Organism An individual living thing.

Peptide Essentially just a very short protein, composed of only a few linked amino acids. The transition point between peptides and proteins is not specifically defined.

Phosphate group Common chemical group consisting of phosphorous and oxygen atoms (and sometimes with some hydrogen atoms attached) which is vital to much of the chemistry of life.

Photosynthesis The process within chloroplasts in which the energy of sunlight is trapped and used to generate ATP and NADPH, and then this ATP and NADPH is used to incorporate atoms from carbon dioxide gas into the complex organic chemicals of the cell.

Polar covalent bond A chemical bond between two atoms which is formed when electrons become shared between the atoms involved, but shared unequally, so that one atom possesses a slight positive charge (δ^+) while the other possesses a slight negative charge (δ^-).

Polarized membrane A membrane with a voltage difference (potential difference) across it, due to some imbalance in the numbers of various ions on either side of the membrane.

Potential energy A form of energy which things possess because their *position* involves some resistance against a fundamental force.

Proteins Giant molecules formed when many individual amino acids become linked together. They catalyse and control almost all the chemical processes of life, as well as performing various structural and other roles.

Proton A sub-atomic particle which carries a positive charge. Found in the nucleus of an atom.

Protozoa Single-celled micro-organisms which, unlike bacteria, contain a nucleus and various other organelles.

Quark Believed to be one of the fundamental types of particles of matter, from which protons, for example, are constructed.

Radiant energy Electromagnetic radiation such as sunlight, radio waves, x-rays, etc.

Receptor proteins Proteins found in cell membranes which selectively bind to chemicals outside of the cell and alter the cell's activity as a result.

Recombination The process of swapping over which allows sections of DNA with similar, but different, base sequences to change places with one another.

Replication of DNA The copying of one original double-helix into two copies of itself.

Ribosome The complex of proteins and RNAs on which protein manufacture takes place.

RNA Ribonucleic acid. The nucleic acid which acts as the intermediary between DNA and protein in the central mechanism of life.

Sexual reproduction Form of reproduction which involves the union of the germ cells of two different organisms, to create the first cell of a new organism.

Space-time The four-dimensional universe formed from the three dimensions of space united with the one dimension of time.

Sperm cell The germ cell of a male sexually reproducing organism.

Splicing The process in which the non-coding introns of an RNA molecule are removed, and the coding portions joined together, to generate a mature messenger RNA molecule.

Stroma The fluid-filled compartment of a chloroplast between the inner membrane and the thylakoid membranes.

Strong nuclear force One of the fundamental forces of nature. It is responsible for holding protons together in the nucleus of an atom, despite the fact that, being positively charged, they would otherwise be repelled from one another by the electromagnetic force.

Substrates The starting materials of an enzyme-catalysed reaction.

Sugars Small carbohydrate molecules.

Synapse The space between two nerve cells, across which neurotransmitters diffuse to allow the activity of one nerve cell to affect the activity of another.

Thermal motion The constant random motion of particles due to their possession of the kinetic energy of heat.

Thylakoid membrane The membrane within a chloroplast which houses the essential mechanisms of photosynthesis, which allow sunlight to be used to generate ATP (and NADPH).

Thylakoid space The space enclosed by the thylakoid membrane of a chloroplast.

Thymine One of the bases found in DNA. It pairs with adenine to form the A-T base-pair.

Tissue A group of cells of the same general type, e.g. muscle tissue, lung tissue, etc.

Tissue factors Small molecules released from one cell in order to bind to and bring about some effect in nearby cells, especially cells of the same tissue.

Transcription The copying of a strand of DNA into a complementary strand of RNA.

Transfer RNAs (tRNAs) The RNA molecules which bring specific amino acids to the ribosome during protein synthesis, and transfer them into a growing protein chain.

Translation The decoding of the genetic information (base sequence) of a messenger RNA into the amino acid sequence of the protein molecule which the messenger RNA encodes.

Transposition A process in which sections of DNA can jump from one region of a genome to another, either by detaching from their original location to become reincorporated elsewhere in the genome, or by giving rise to copies which move to new locations.

Tricarboxylic acid cycle (Citric acid cycle, Kreb's cycle) The metabolic cycle in the mitochondrion in which pyruvate molecules are used as a source of the energy needed to generate ATP.

Uracil One of the bases found in RNA. It pairs with adenine to form the A-U base-pair.

Van der Waals forces (Van der Waals bonds) Forces of attraction between chemicals, caused by transient fluctuating partial charges on their surface created by the random motion of their electrons.

Vesicle Any small sac within the cell bounded by a lipid bilayer membrane.

Weak nuclear force One of the fundamental forces of nature. It is responsible for some forms of radioactive decay within atomic nuclei. It is probably just a subtle additional manifestation of the electromagnetic force.

Further reading

Here is a short list of some good books which will provide you with more details of the topics introduced to you in this one. Most of these books are written with the needs of the novice in mind, apart from the more conventional textbooks marked with an asterisk.

Fundamental physics

The Cosmic Onion, by Frank Close, Heinemann, 1983.
The Forces of Nature, by Paul Davies, Cambridge University Press, 1979.
Superforce, by Paul Davies, Unwin, 1984.
The Cosmic Code, by Heinz R. Pagels, Penguin, 1982.
Time, Space and Things, by B. K. Ridley, Cambridge University Press, 1984.

Chemical principles

The Second Law, by Peter W. Atkins, W. H. Freeman, 1984.
Chemical Principles *, by William L. Masterton, Emil J. Slowinski and Conrad L. Stanitski, Saunders College Publishing, 1985.
Introducing Chemistry, by Hazel Rossotti, Penguin, 1975.
The Marvels of the Molecule, by Lionel Salem, VCH (W. Germany), 1987.

Biochemistry and molecular biology

*Molecular Biology of the Cell**, by Bruce Alberts, Dennis Bray, Julian Lewis, Martin Raff, Keith Roberts and James D. Watson, Garland Publishing, 1983.

A Guided Tour of the Living Cell, by Christian de Duve, Scientific American Books, 1984.

The Cartoon Guide to Genetics, by Larry Gonick and Mark Wheelis, Barnes & Noble, 1983.

In Search of the Double Helix, by John Gribbin, Corgi, 1985.

The Chemistry of Life, by Stephen Rose, Penguin, 1979.

The nervous system

Mindwaves, edited by Colin Blakemore and Susan Greenfield, Basil Blackwell, 1987.

The Oxford Companion to the Mind, edited by Richard. L. Gregory, Oxford University Press, 1987.

*Neurobiology**, by Gordon M. Shepherd, Oxford University Press, 1983.

Life's origin and evolution

The Selfish Gene, by Richard Dawkins, Paladin, 1976.

The Blind Watchmaker, by Richard Dawkins, Longman, 1986.

The Theory of Evolution, by John Maynard Smith, Penguin, 1975.

Darwin for Beginners, by Jonathan Miller, Unwin, 1982.

The Creation of Life, by Andrew Scott, Basil Blackwell, 1986.

Index

Vital Principles
The Molecular Mechanisms of Life
Andrew Scott

Each day the earth spins in the radiant energy
of the sun and something remarkable happens.
Some of the energy is captured within living
cells to power the creation of new life from the
lifeless minerals of the world. In a feat of
stunning self-regulating choreography, billions
of atoms, molecules and ions become a part of
the frantic chemical dance we call life. If we
examine living things to discover what they are
made of and how they work, we find that they
are intricate chemical machines composed of
and sustained by chemicals interacting
according to the fundamental laws of physics.
The continual dynamic interplay of chemicals as
they react to one another's presence is all we
can find within us.

Andrew Scott describes the essential physics,
chemistry and biology of life. He reveals what
living things are made of and how they work,
beginning at the level of atoms and then
working up through genes and proteins to the
cells, tissues and organs of multicellular life.
This 'quick guide' to the central principles of
the living world explores modern science's
ideas about the essence of life, ideas which
replace the mysterious 'vital principles' with
which living matter was once thought to be
endowed.

Vital Principles offers its readers an easy
insight into the fascinating microworld that
makes us what we are.